AF664162

PLANE GEOMETRY

G.A. WENTWORTH
Author of a Series of Text-Books in Mathematics

HAWK PRESS

Published by

Hawk Press
4836/24, Ansari Road, Daryaganj
New Delhi - 110 002
Phones : 9643330713, 011-23278618, 011-35676207
E-mail : thehawkpress@gmail.com
www.thehawkpress.com

Copyright © 2022, Editor
ISBN : 978-93-93971-44-9
All rights reserved.

No part of this book may be reproduced, stored in a retrieval system, transmitted or utilised in any form or by any means, electronic, mechanical, photocopying, recording or otherwise, without the prior permission of the copyright owner. Application for such permission should be addressed to the publisher.

PREFACE.

Most persons do not possess, and do not easily acquire, the power of abstraction requisite for apprehending geometrical conceptions, and for keeping in mind the successive steps of a continuous argument. Hence, with a very large proportion of beginners in Geometry, it depends mainly upon the *form* in which the subject is presented whether they pursue the study with indifference, not to say aversion, or with increasing interest and pleasure.

Great care, therefore, has been taken to make the pages attractive. The figures have been carefully drawn and placed in the middle of the page, so that they fall directly under the eye in immediate connection with the text; and in no case is it necessary to turn the page in reading a demonstration. Full, long-dashed, and short-dashed lines of the figures indicate given, resulting, and auxiliary lines, respectively. Bold-faced, italic, and roman type has been skilfully used to distinguish the hypothesis, the conclusion to be proved, and the proof.

As a further concession to the beginner, the reason for each statement in the early proofs is printed in small italics, immediately following the statement. This prevents the necessity of interrupting the logical train of thought by turning to a previous section, and compels the learner to become familiar with a large number of geometrical truths by constantly seeing and repeating them. This help is gradually discarded, and the pupil is left to depend upon the knowledge already acquired, or to find the reason for a step by turning to the given reference.

It must not be inferred, because this is not a geometry of interrogation points, that the author has lost sight of the real object of the study. The training to be obtained from carefully following the logical steps of a complete proof has been provided for by the Propositions of the Geometry, and the development of the power to grasp and prove new truths has been provided for by original exercises. The chief value of any Geometry consists in the happy combination of these two kinds of training. The exercises have been arranged according to the test of experience, and are so abundant that it is not expected that any one class will work them all out. The methods of attacking and proving original theorems are fully explained in the first Book, and illustrated by sufficient examples; and the methods of attacking and solving original problems are explained in the second Book, and illustrated by examples worked out in full. None but the very simplest exercises are inserted until the student has become familiar with geometrical methods, and is furnished with elementary but much needed instruction in the art of handling original propositions; and he is assisted by diagrams and hints as long as these helps are necessary to develop his mental powers sufficiently to enable him to carry on the work by himself.

The law of converse theorems, the distinction between positive and negative quantities, and the principles of reciprocity and continuity have been briefly explained; but the application of these principles is left mainly to the discretion of teachers.

The author desires to express his appreciation of the valuable suggestions and assistance which he has received from distinguished educators in all parts of the country. He also desires to acknowledge his obligation to Mr. Charles Hamilton, the Superintendent of the composition room of the Athenæum Press, and to Mr. I. F. White, the compositor, for the excellent typography of the book.

Criticisms and corrections will be thankfully received.

G. A. WENTWORTH.

EXETER, N.H., June, 1899.

NOTE TO TEACHERS.

It is intended to have the first sixteen pages of this book simply read in the class, with such running comment and discussion as may be useful to help the beginner catch the spirit of the subject-matter, and not leave him to the mere letter of dry definitions. In like manner, the definitions at the beginning of each Book should be read and discussed in the recitation room. There is a decided advantage in having the definitions for each Book in a single group so that they can be included in one survey and discussion.

For a similar reason the theorems of limits are considered together. The subject of limits is exceedingly interesting in itself, and it was thought best to include in the theory of limits in the second Book every principle required for Plane and Solid Geometry.

When the pupil is reading each Book for the first time, it will be well to let him write his proofs on the blackboard in his own language, care being taken that his language be the simplest possible, that the arrangement of work be vertical, and that the figures be accurately constructed.

This method will furnish a valuable exercise as a language lesson, will cultivate the habit of neat and orderly arrangement of work, and will allow a brief interval for deliberating on each step.

After a Book has been read in this way, the pupil should review the Book, and should be required to draw the figures free-hand. He should state and prove the propositions orally, using a pointer to indicate on the figure every line and angle named. He should be encouraged in reviewing each Book, to do the original exercises; to state the converse propositions, and determine whether they are true or false; and also to give well-considered answers to questions which may be asked him on many propositions.

The Teacher is strongly advised to illustrate, geometrically and arithmetically, the principles of limits. Thus, a rectangle with a constant base b, and a variable altitude x, will afford an obvious illustration of the truth that the product of a constant and a variable is also a variable; and that the limit of the product of a constant and a variable is the product of the constant by the limit of the variable. If x increases and approaches the altitude a as a limit, the area of the rectangle increases and approaches the area of the rectangle ab as a limit; if, however, x decreases and approaches zero as a limit, the area of the rectangle decreases and approaches zero as a limit.

An arithmetical illustration of this truth may be given by multiplying the approximate values of any repetend by a constant. If, for example, we take the repetend 0.3333 etc., the approximate values of the repetend will be $\frac{3}{10}$, $\frac{33}{100}$, $\frac{333}{1000}$, $\frac{3333}{10000}$, etc., and these values multiplied by 60 give the series 18, 19.8, 19.98, 19.998, etc., which evidently approaches 20 as a limit; but the product of 60 into $\frac{1}{3}$ (the limit of the repetend 0.333 etc.) is also 20.

Again, if we multiply 60 into the different values of the decreasing series $\frac{1}{30}$, $\frac{1}{300}$, $\frac{1}{3000}$, $\frac{1}{30000}$, etc., which approaches zero as a limit, we shall get the decreasing series 2, $\frac{1}{5}$, $\frac{1}{50}$, $\frac{1}{500}$, etc.; and this series evidently approaches zero as a limit.

The Teacher is likewise advised to give frequent written examinations. These should not be too difficult, and sufficient time should be allowed for accurately constructing the figures, for choosing the best language, and for determining the best arrangement.

The time necessary for the reading of examination books will be diminished by more than one half, if the use of symbols is allowed.

EXETER, N.H., 1899.

Contents

GEOMETRY.

INTRODUCTION.

1. If a block of wood or stone is cut in the shape represented in Fig. 1, it will have *six flat faces.*

Each face of the block is called a surface; and if the faces are made smooth by polishing, so that, when a straight edge is applied to any one of them, the straight edge in every part will touch the surface, the faces are called **plane surfaces**, or **planes**.

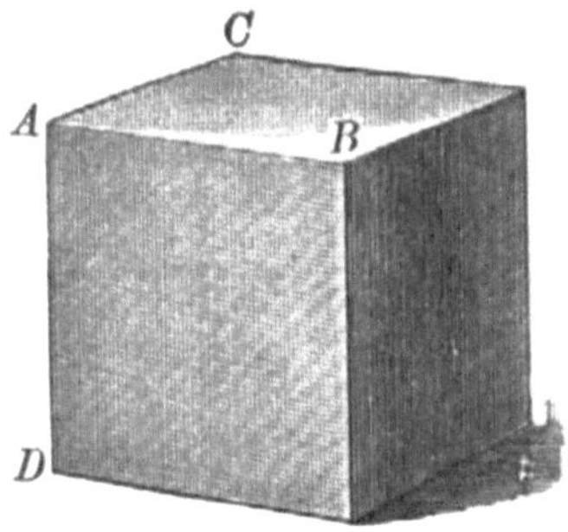

FIG. 1.

2. The intersection of any two of these surfaces is called a **line**.

3. The intersection of any three of these lines is called a **point**.

4. The block extends in three principal directions:

From left to right, A to B.

From front to back, A to C.

From top to bottom, A to D.

These are called the **dimensions** of the block, and are named in the order given, **length**, **breadth** (or *width*), and **thickness** (*height* or *depth*).

5. A **solid**, in common language, is a limited portion of space *filled with matter*; but in Geometry we have nothing to do with *the matter* of which a body is composed; we study simply its *shape* and *size*; that is, we regard a solid as a limited portion of space which may be occupied by a physical body, or marked out in some other way. Hence,

A geometrical solid is a limited portion of space.

6. The surface of a solid is simply the boundary of the solid, that which separates it from surrounding space. The surface is no part of a solid and has no thickness. Hence,

A surface has only two dimensions, length and breadth.

7. A line is simply a boundary of a surface, or the intersection of two surfaces. Since the surfaces have no thickness, a line has no thickness. Moreover, a line is no part of a surface and has no width. Hence,

A line has only one dimension, length.

8. A point is simply the extremity of a line, or the intersection of two lines. A point, therefore, has no thickness, width, or length; therefore, no magnitude. Hence,

A point has no dimension, but denotes position simply.

9. It must be distinctly understood at the outset that the points, lines, surfaces, and solids of Geometry are *purely ideal*, though they are represented to the eye in a material way. Lines, for example, drawn on paper or on the blackboard, will have some width and some thickness, and will so far fail of being *true lines*; yet, when they are used to help the mind in reasoning, it is assumed that they represent true lines, without breadth and without thickness.

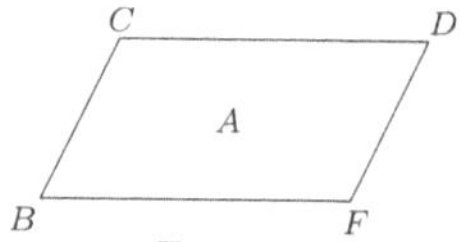

FIG. 2.

10. A point is *represented* to the eye by a fine dot, and named by a letter, as A (Fig. 2). A line is named by two letters, placed one at each end, as BF. A surface is represented and named by the lines which bound it, as $BCDF$. A solid is represented by the faces which bound it.

11. A point in space may be considered by itself, without reference to a line.

12. If a point moves in space, its path is a line. This line may be considered apart from the idea of a surface.

13. If a line moves in space, it generates, in general, a surface. A surface can then be considered apart from the idea of a solid.

14. If a surface moves in space, it generates, in general, a solid.

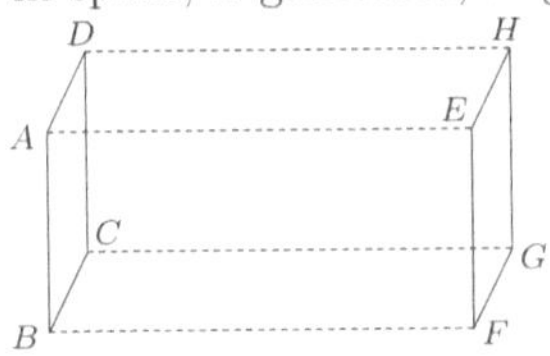

FIG. 3.

Thus, let the upright surface $ABCD$ (Fig. 3) move to the right to the position $EFGH$, the points A, B, C, and D generating the lines AE, BF, CG, and DH, respectively. The lines AB, BC, CD, and DA will generate the surfaces AF. BG, CH, and DE, respectively. The surface $ABCD$ will generate the solid AG.

15. Geometry is the science which treats of *position, form*, and *magnitude.*

16. A **geometrical figure** is a combination of points, lines, surfaces, or solids.

17. Plane Geometry treats of figures all points of which are in the same plane.

Solid Geometry treats of figures all points of which are not in the same plane.

GENERAL TERMS.

18. A **proof** is a course of reasoning by which the truth or falsity of any statement is logically established.

19. An **axiom** is a statement admitted to be true without proof.

20. A **theorem** is a statement to be proved.

21. A **construction** is the representation of a required figure by means of points and lines.

22. A **postulate** is a construction admitted to be possible.

23. A **problem** is a construction to be made so that it shall satisfy certain given conditions.

24. A **proposition** is an axiom, a theorem, a postulate, or a problem.

25. A **corollary** is a truth that is easily deduced from known truths.

26. A **scholium** is a remark upon some particular feature of a proposition.

27. The **solution of a problem** consists of four parts:

1. The *analysis*, or course of thought by which the construction of the required figure is discovered.
2. The *construction* of the figure with the aid of ruler and compasses.
3. The *proof* that the figure satisfies all the conditions.
4. The *discussion* of the limitations, if any, within which the solution is possible.

28. A theorem consists of two parts: the **hypothesis**, or that which is assumed; and the **conclusion**, or that which is asserted to follow from the hypothesis.

29. The **contradictory** of a theorem is a theorem which must be true if the given theorem is false, and must be false if the given theorem is true. Thus,

A theorem:	If A is B, then C is D.
Its contradictory:	If A is B, then C is not D.

30. The **opposite** of a theorem is obtained by making *both the hypothesis and the conclusion negative.* Thus,

A theorem:	If A is B, then C is D.
Its opposite:	If A is not B, then C is not D.

31. The **converse** of a theorem is obtained by *interchanging the hypothesis and conclusion.* Thus,

A theorem:	If A is B, then C is D.
Its converse:	If C is D, then A is B.

32. The converse of a truth is not *necessarily* true.

Thus, Every horse is a quadruped is true, but the converse, Every quadruped is a horse, is not true.

33. *If a direct proposition and its opposite are true, the converse proposition is true; and if a direct proposition and its converse are true, the opposite proposition is true.*

Thus, if it were true that

1. If an animal is a horse, the animal is a quadruped;
2. If an animal is not a horse, the animal is not a quadruped;

it would follow that

3. If an animal is a quadruped, the animal is a horse.

Moreover, if 1 and 3 were true, then 2 would be true.

34. GENERAL AXIOMS.

1. Magnitudes which are equal to the same magnitude, or equal magnitudes, are equal to each other.

2. If equals are added to equals, the sums are equal.

3. If equals are taken from equals, the remainders are equal.

4. If equals are added to unequals, the sums are unequal in the same order; if unequals are added to unequals in the same order, the sums are unequal in that order.

5. If equals are taken from unequals, the remainders are unequal in the same order; if unequals are taken from equals, the remainders are unequal in the reverse order.

6. The doubles of the same magnitude, or of equal magnitudes are equal; and the doubles of unequals are unequal.

7. The halves of the same magnitude, or of equal magnitudes are equal; and the halves of unequals are unequal.

8. The whole is greater than any of its parts.

9. The whole is equal to the sum of all its parts.

35. SYMBOLS AND ABBREVIATIONS.

$>$	is (or are) greater than.	Def.	...definition.
$<$	is (or are) less than.	Ax.	...axiom.
$\backsimeq$	is (or are) equivalent to.	Hyp.	...hypothesis.
$\therefore$	therefore.	Cor.	...corollary.
$\perp$	perpendicular.	Scho.	...scholium.
$\perp_s$	perpendiculars.	Ex.	...exercise.
$\parallel$	parallel. $\parallel_s$ parallels.	Adj.	...adjacent.
$\angle$	angle. $\angle_s$ angles.	Iden.	...identical.
$\triangle$	triangle. $\triangle_s$ triangles.	Const.	...construction.
▱	parallelogram.	Sup.	...supplementary.
▱$_s$	parallelograms.	Ext.	...exterior.
$\odot$	circle. $\odot_s$ circles.	Int.	...interior.
rt.	right. st. straight.	Alt.	...alternate.

Q.E.D. stands for quod erat demonstrandum, *which was to be proved.*

Q.E.F. stands for quod erat faciendum, *which was to be done.*

The signs $+$, $-$, $\times$, $\div$, $=$, have the same meaning as in Algebra.

PLANE GEOMETRY.

BOOK I. RECTILINEAR FIGURES.

DEFINITIONS.

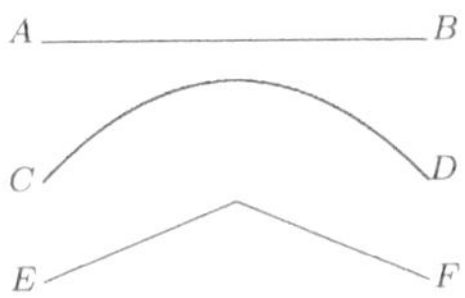

FIG. 4.

36. A **straight line** is a line such that any part of it, however placed on any other part, will lie wholly in that part if its extremities lie in that part, as AB.

37. A **curved line** is a line no part of which is straight, as CD.

38. A **broken line** is made up of different straight lines, as EF.

NOTE. A straight line is often called simply a *line*.

39. A **plane surface**, or a **plane**, is a surface in which, if any two points are taken, the straight line joining these points lies wholly in the surface.

40. A **curved surface** is a surface no part of which is plane.

41. A **plane figure** is a figure all points of which are in the same plane.

42. Plane figures which are bounded by straight lines are called **rectilinear** figures; by curved lines, **curvilinear** figures.

43. Figures that have the *same shape* are called **similar**. Figures that have the *same size but not the same shape* are called **equivalent**. Figures that have the *same shape and the same size* are called **equal** or **congruent**.

THE STRAIGHT LINE.

44. Postulate. A straight line can be drawn from one point to another.

45. Postulate. A straight line can be produced indefinitely.

46. Axiom.* *Only one straight line can be drawn from one point to another.* Hence, two points *determine* a straight line.

47. COR. 1. *Two straight lines which have two points in common coincide and form but one line.*

48. COR. 2. *Two straight lines can intersect in only one point.*

For if they had two points common, they would coincide and not intersect.

Hence, two intersecting lines *determine* a point.

49. Axiom. *A straight line is the shortest line that can be drawn from one point to another.*

50. DEF. The **distance** between two points is the length of the straight line that joins them.

51. A straight line determined by two points may be considered as prolonged indefinitely.

52. If only the part of the line between two fixed points is considered, this part is called a **segment** of the line.

53. For brevity, we say "the line AB," to designate a segment of a line limited by the points A and B.

54. If a line is considered as extending from a fixed point, this point is called the **origin** of the line.

*The general axioms on page 6 apply to all magnitudes. Special geometrical axioms will be given when required.

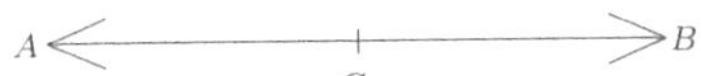

FIG. 5.

55. If any point, C, is taken in a given straight line, AB, the two parts CA and CB are said to have *opposite directions* from the point C (Fig. 5).

Every straight line, as AB, may be considered as extending in either of two opposite directions, namely, from A towards B, which is expressed by AB, and read segment AB; and from B towards A, which is expressed by BA, and read segment BA.

56. If the magnitude of a given line is changed, it becomes longer or shorter.

Thus (Fig. 5), by prolonging AC to B we add CB to AC, and $AB = AC + CB$. By diminishing AB to C, we subtract CB from AB, and $AC = AB - CB$.

If a given line increases so that it is prolonged by its own magnitude several times in succession, the line is *multiplied*, and the resulting line is called a *multiple* of the given line.

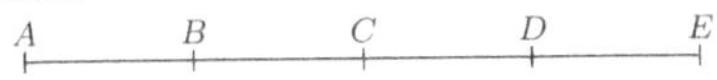

FIG. 6.

Thus (Fig. 6), if $AB = BC = CD = DE$, then $AC = 2AB$, $AD = 3AB$, and $AE = 4AB$. Hence,

Lines of given length may be added and subtracted; they may also be multiplied by a number.

THE PLANE ANGLE.

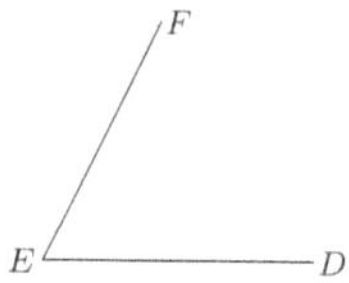

FIG. 7.

57. The *opening* between two straight lines drawn from the same point is called a **plane angle**. The two lines, ED and EF, are called the **sides**, and E, the point of meeting, is called the **vertex** of the angle.

The size of an angle depends upon the *extent of opening* of its sides, and not upon the length of its sides.

58. If there is but one angle at a given vertex, the angle is designated by a capital letter placed at the vertex, and is read by simply naming the letter.

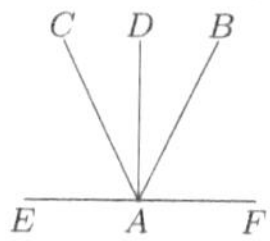

FIG. 8.

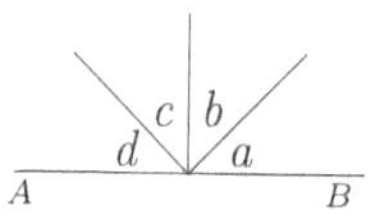

FIG. 9.

If two or more angles have the same vertex, each angle is designated by three letters, and is read by naming the three letters, the one at the vertex between the others. Thus, DAC (Fig. 8) is the angle formed by the sides AD and AC.

An angle is often designated by placing a small *italic* letter between the sides and near the vertex, as in Fig. 9.

59. Postulate of Superposition. Any figure may be moved from one place to another without altering its size or shape.

60. The **test of equality** of two geometrical magnitudes is that they may be made to coincide throughout their whole extent. Thus,

Two straight lines are equal, if they can be placed one upon the other so that the points at their extremities coincide.

Two angles are equal, if they can be placed one upon the other so that their vertices coincide and their sides coincide, each with each.

61. A line or plane that divides a geometric magnitude into *two equal parts* is called the **bisector** of the magnitude.

If the angles BAD and CAD (Fig. 8) are equal, AD *bisects* the angle BAC.

62. Two angles are called **adjacent angles** when they have the same vertex and a common side between them; as the angles BOD and AOD (Fig. 10).

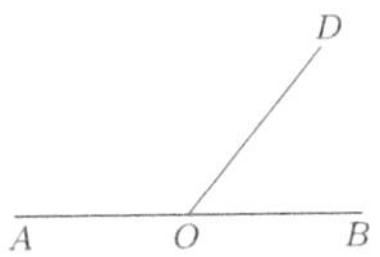

FIG. 10.

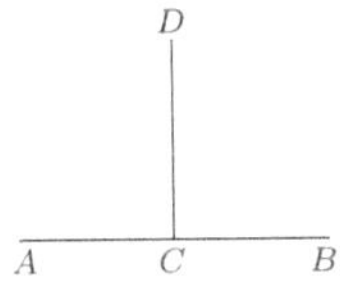

FIG. 11.

63. When one straight line meets another straight line and makes the *adjacent angles equal*, each of these angles is called a **right angle**; as angles DCA and DCB (Fig. 11).

64. A **perpendicular** to a straight line is a straight line that makes a right angle with it.

Thus, if the angle DCA (Fig. 11) is a right angle, DC is perpendicular to AB, and AB is perpendicular to DC.

65. The point (as C, Fig. 11) where a perpendicular meets another line is called the **foot** of the perpendicular.

66. When the sides of an angle extend in opposite directions, so as to be in the same straight line, the angle is called a **straight angle**.

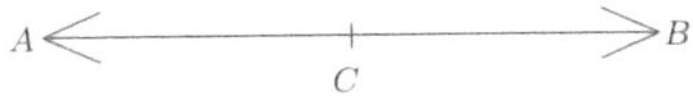

FIG. 12.

Thus, the angle formed at C (Fig. 12) with its sides CA and CB extending in opposite directions from C is a straight angle.

67. COR. *A right angle is half a straight angle.*

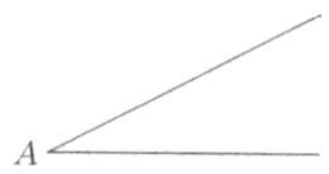

FIG. 13.

FIG. 14.

68. An angle less than a right angle is called an **acute angle**; as, angle A (Fig. 13).

69. An angle greater than a right angle and less than a straight angle is called an **obtuse angle**; as, angle AOD (Fig. 14).

70. An angle greater than a straight angle and less than two straight angles is called a **reflex angle**; as, angle DOA, indicated by the dotted line (Fig. 14).

71. Angles that are neither right nor straight angles are called **oblique angles**; and intersecting lines that are not perpendicular to each other are called **oblique lines**.

EXTENSION OF THE MEANING OF ANGLES.

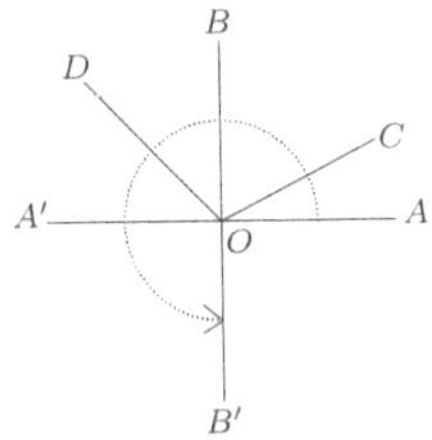

FIG. 15.

72. Suppose the straight line OC (Fig. 15) to move in the plane of the paper from coincidence with OA, about the point O as a pivot, to the position OC; then the line OC describes or generates *the angle* AOC, and the magnitude of the angle AOC depends upon the *amount of rotation* of the line from the position OA to the position OC.

If the rotating line moves from the position OA to the position OB, *perpendicular* to OA, it generates the right angle AOB; if it moves to the position OD, it generates the obtuse angle AOD; if it moves to the position OA', it generates the straight angle AOA'; if it moves to the position OB' it generates the reflex angle AOB', indicated by the dotted line; and if it moves to the position OA again, it generates two straight angles. Hence,

73. *The angular magnitude about a point in a plane is equal to two straight angles, or four right angles; and the angular magnitude about a point on one side of a straight line drawn through the point is equal to a straight angle, or two right angles.*

74. The whole angular magnitude about a point in a plane is called a **perigon**; and two angles whose sum is a perigon are called **conjugate angles**.

NOTE. This *extension of the meaning of angles* is necessary in the applications of Geometry, as in Trigonometry, Mechanics, etc.

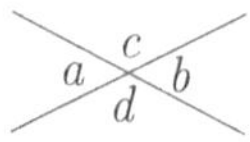

FIG. 16.

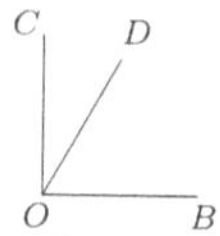

FIG. 17.

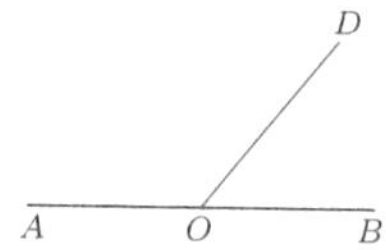

FIG. 18.

75. When two angles have the same vertex, and the sides of the one are prolongations of the sides of the other, they are called **vertical angles**; as, angles a and b, c and d (Fig. 16).

76. Two angles are called **complementary** when their sum is equal to a right angle; and each is called the *complement* of the other; as, angles DOB and DOC (Fig. 17).

77. Two angles are called **supplementary** when their sum is equal to a straight angle; and each is called the *supplement* of the other; as, angles DOB and DOA (Fig. 18).

UNIT OF ANGLES.

78. By adopting a suitable unit for measuring angles we are able to express the magnitudes of angles by numbers.

If we suppose OC (Fig. 15) to turn about O from coincidence with OA until it makes *one three hundred sixtieth* of a revolution, it generates an angle at O, which is taken as the unit for measuring angles. This unit is called a *degree.*

The degree is subdivided into sixty equal parts, called *minutes*, and the minute into sixty equal parts, called *seconds.*

Degrees, minutes, and seconds are denoted by symbols. Thus, 5 degrees 13 minutes 12 seconds is written $5° \ 13' \ 12''$.

A right angle is generated when OC has made *one fourth* of a revolution and contains 90°; a straight angle, when OC has made *half* of a revolution and contains 180°; and a perigon, when OC has made a complete revolution and contains 360°.

NOTE. The natural angular unit is one complete revolution. But this unit would require us to express the values of most angles by fractions. The advantage of using the degree as the unit consists in its convenient size, and in the fact that 360 is divisible by so many different integral numbers.

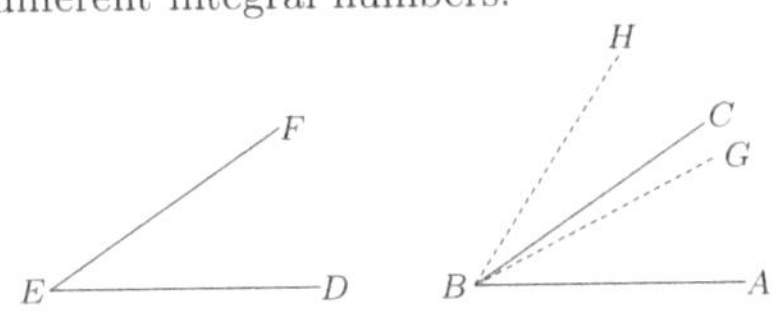

FIG. 19.

79. By the method of superposition we are able to compare magnitudes of the same kind. Suppose we have two angles, ABC and DEF (Fig. 19). Let the side ED be placed on the side BA, so that the vertex E shall fall on B; then, if the side EF falls on BC, the angle DEF equals the angle ABC; if the side EF falls between BC and BA in the position shown by the dotted line BG, the angle DEF is less than the angle ABC; but if the side EF falls in the position shown by the dotted line BH, the angle DEF is greater than the angle ABC.

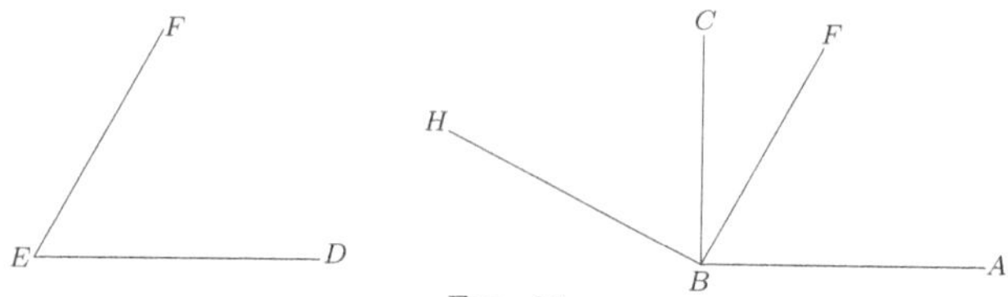

FIG. 20.

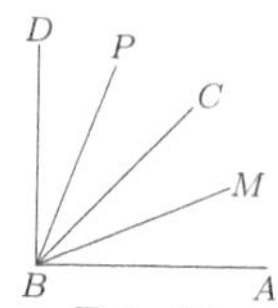

FIG. 21.

80. If we have the angles ABC and DEF (Fig. 20), and place the vertex E on B and the side ED on BC, so that the angle DEF takes the position CBH, the angles DEF and ABC will together be equal to the angle ABH.

If the vertex E is placed on B, and the side ED on BA, so that the angle DEF takes the position ABF, the angle FBC will be the difference between the angles ABC and DEF.

If an angle is increased by its own magnitude two or more times in succession, the angle is *multiplied* by a number.

Thus, if the angles ABM, MBC, CBP, PBD (Fig. 21) are all equal, the angle ABD is 4 times the angle ABM. Therefore,

Angles may be added and subtracted; they may also be multiplied by a number.

PERPENDICULAR AND OBLIQUE LINES.

PROPOSITION I. THEOREM.

81. *All straight angles are equal.*

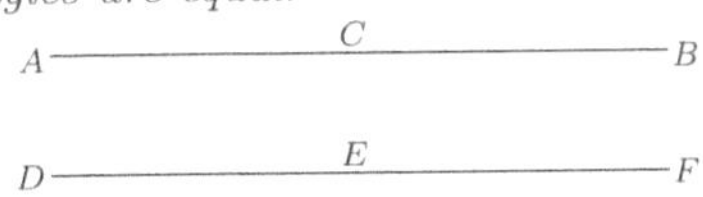

Let the angles ACB and DEF be any two straight angles.

To prove that $\angle ACB = \angle DEF$.

Proof. Place the $\angle ACB$ on the $\angle DEF$, so that the vertex C shall fall on the vertex E, and the side CB on the side EF.

Then CA will fall on ED, § 47

(*because ACB and DEF are straight lines*).

$\therefore \angle ACB = \angle DEF$. § 60

Q.E.D.

82. COR. 1. *All right angles are equal.* Ax. 7

83. COR. 2. *At a given point in a given line there can be but one perpendicular to the line.*

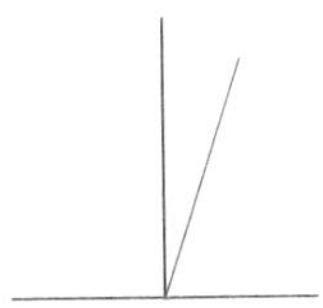

For, if there could be two $\perp_s$, we should have rt. $\angle_s$ of different magnitudes; but this is impossible, § 82.

84. COR. 3. *The complements of the same angle or of equal angles are equal.* Ax. 3

85. COR. 4. *The supplements of the same angle or of equal angles are equal.* Ax. 3

NOTE. The beginner must not forget that in Plane Geometry all the points of a figure are in the same plane. Without this restriction in Cor. 2, an indefinite number of perpendiculars can be erected at a given point in a given line.

PROPOSITION II. THEOREM.

86. *If two adjacent angles have their exterior sides in a straight line, these angles are supplementary.*

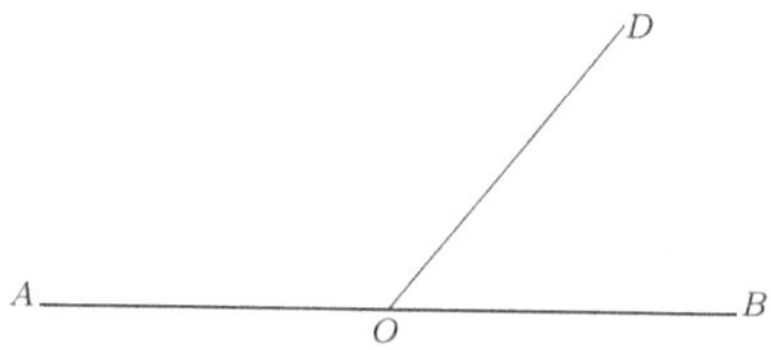

Let the exterior sides OA and OB of the adjacent angles AOD and BOD be in the straight line AB.

To prove that $\angle_s AOD$ *and* BOD *are supplementary.*

Proof.

AOB is a straight line. Hyp.

$\therefore \angle AOB$ is a st. $\angle$. § 66

But

$\angle AOD + \angle BOD =$ the st. $\angle AOB$. Ax. 9

$\therefore$ the $\angle_s AOD$ and BOD are supplementary. § 77

Q.E.D.

87. DEF. Adjacent angles that are supplements of each other are called *supplementary-adjacent angles.*

Since the angular magnitude about a point is neither increased nor diminished by the number of lines which radiate from the point, it follows that,

88. COR. 1. *The sum of all the angles about a point in a plane is equal to a perigon, or two straight angles.*

89. COR. 2. *The sum of all the angles about a point in a plane, on the same side of a straight line passing through the point, is equal to a straight angle, or two right angles.*

PROPOSITION III. THEOREM.

90. CONVERSELY: *If two adjacent angles are supplementary, their exterior sides are in the same straight line.*

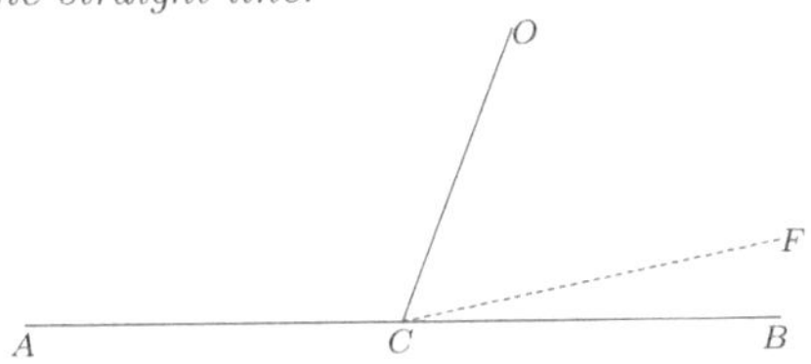

Let the adjacent angles OCA and OCB be supplementary.

To prove that AC and CB are in the same straight line.

Proof. Suppose CF to be in the same line with AC.

Then

$\angle_s OCA$ and OCF are supplementary, § 86

(*if two adjacent angles have their exterior sides in a straight line, these angles are supplementary*).

But

$\angle_s OCA$ and OCB are supplementary. Hyp.

$\therefore \angle_s OCF$ and OCB have the same supplement.

$\therefore \angle OCF = \angle OCB$. § 85

$\therefore CB$ and CF coincide. § 60

$\therefore AC$ and CB are in the same straight line. Q.E.D.

Since Propositions II. and III. are true, their opposites are true. Hence, § 33

91. COR. 1. *If the exterior sides of two adjacent angles are not in a straight line, these angles are not supplementary.*

92. COR. 2. *If two adjacent angles are not supplementary, their exterior sides are not in the same straight line.*

Proposition IV. Theorem.

93. *If one straight line intersects another straight line, the vertical angles are equal.*

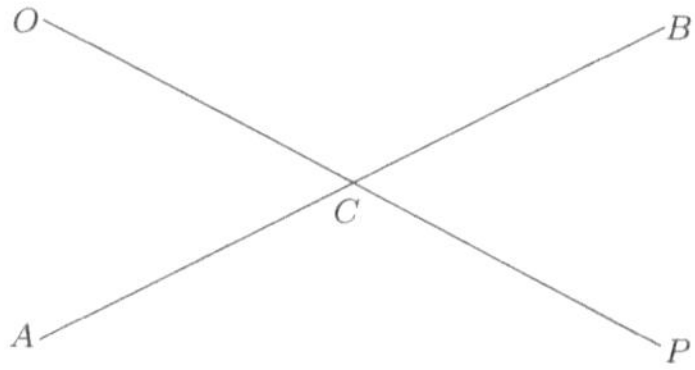

Let the lines OP and AB intersect at C.
To prove that

$$\angle OCB = \angle ACP.$$

Proof.

$\angle OCA$ and $\angle OCB$ are supplementary. § 86

$\angle OCA$ and $\angle ACP$ are supplementary, § 86

(*if two adjacent angles have their exterior sides in a straight line, these angles are supplementary*).

$\therefore \angle_s OCB$ and ACP have the same supplement.

$\therefore \angle OCB = \angle ACP.$ § 85

In like manner,

$\angle ACO = \angle PCB.$ Q.E.D.

94. Cor. *If one of the four angles formed by the intersection of two straight lines is a right angle, the other three angles are right angles.*

Ex. 1. Find the complement and the supplement of an angle of 49°.

Ex. 2. Find the number of degrees in an angle if it is double its complement; if it is one fourth of its complement.

Ex. 3. Find the number of degrees in an angle if it is double its supplement; if it is one third of its supplement.

PROPOSITION V. THEOREM

95. *Two straight lines drawn from a point in a perpendicular to a given line, cutting off on the given line equal segments from the foot of the perpendicular, are equal and make equal angles with the perpendicular.*

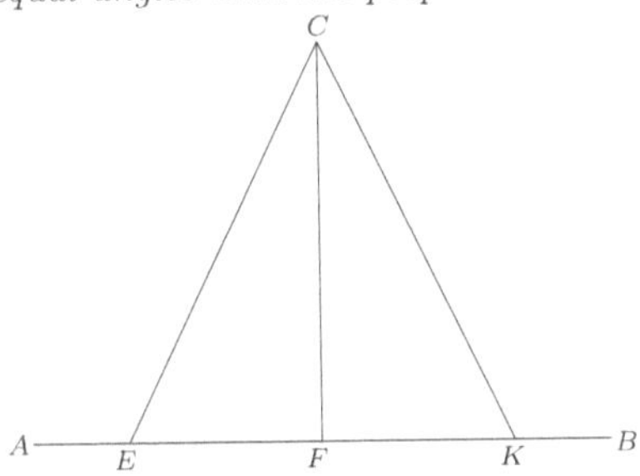

Let CF be a perpendicular to the line AB, and CE and CK two straight lines cutting off on AB equal segments FE and FK from F.

To prove that $CE = CK$; *and* $\angle FCE = \angle FCK$.

Proof. Fold over CFA, on CF as an axis, until it falls on the plane at the right of CF.

FA will fall along FB,
(*since* $\angle CFA = \angle CFB$, *each being a rt.* $\angle$, *by hyp.*).

Point E will fall on point K,
(*since* $FE = FK$, *by hyp.*).

$\therefore CE = CK$, § 60
(*their extremities being the same points*);

and $\angle FCE = \angle FCK$, § 60
(*since their vertices coincide, and their sides coincide, each with each*).

Q.E.D.

Ex. 4. Find the number of degrees in the angle included by the hands of a clock at 1 o'clock. 3 o'clock. 4 o'clock. 6 o'clock.

PROPOSITION VI. THEOREM.

96. *Only one perpendicular can be drawn to a given line from a given external point.*

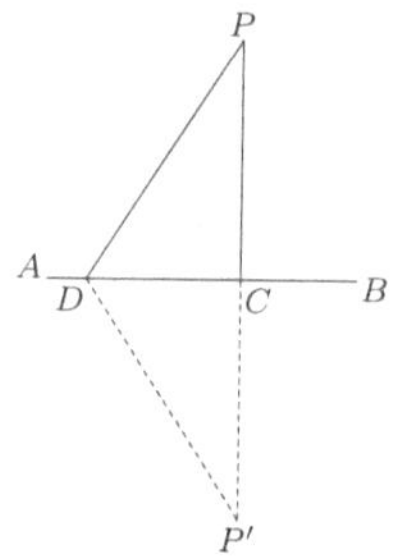

Let AB be the given line, P the given external point, PC a perpendicular to AB from P, and PD any other line from P to AB.

To prove that

PD is not $\perp$ to AB.

Proof. Produce PC to P', making CP' equal to PC.

Draw DP'.

By construction, PCP' is a straight line.

$\therefore PDP'$ is not a straight line, § 46

(*only one straight line can be drawn from one point to another*).

Hence, $\angle PDP'$ is not a straight angle.

Since PC is $\perp$ to DC, and $PC = CP'$,

AC is $\perp$ to PP' at its middle point.

$\therefore \angle PDC = \angle P'DC$, § 95

(*two straight lines from a point in a $\perp$ to a line, cutting off on the line equal segments from the foot of the $\perp$, make equal $\angle_s$ with the $\perp$*)

Since $\angle PDP'$ is not a straight angle,

$\angle PDC$, the half of $\angle PDP'$, is not a right angle.

$\therefore PD$ is not $\perp$ to AB. Q.E.D.

PROPOSITION VII. THEOREM.

97. *The perpendicular is the shortest line that can be drawn to a straight line from an external point.*

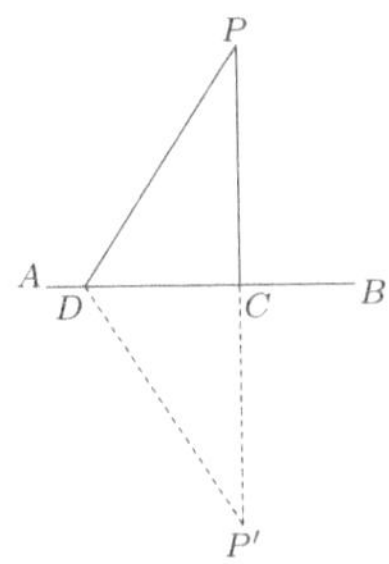

Let AB be the given straight line, P the given point, PC the perpendicular, and PD any other line drawn from P to AB.

To prove that

$$PC < PD.$$

Proof. Produce PC to P', making $CP' = PC$.

Draw DP'.

Then

$$PD = DP', \qquad \text{§ 95}$$

(*two straight lines drawn from a point in a* $\perp$ *to a line, cutting off on the line equal segments from the foot of the* $\perp$, *are equal*).

$$\therefore PD + DP' = 2PD,$$

and

$$PC + CP' = 2PC. \qquad \text{Const.}$$

But

$$PC + CP' < PD + DP'. \qquad \text{§ 49}$$

$$\therefore 2PC < 2PD.$$

$$\therefore PC < PD. \qquad \text{Ax. 7}$$

Q.E.D.

98. COR. *The shortest line that can be drawn from a point to a given line is perpendicular to the given line.*

99. DEF. The **distance** of a point from a line is the length of the perpendicular from the point to the line.

PROPOSITION VIII. THEOREM.

100. *The sum of two lines drawn from a point to the extremities of a straight line is greater than the sum of two other lines similarly drawn, but included by them.*

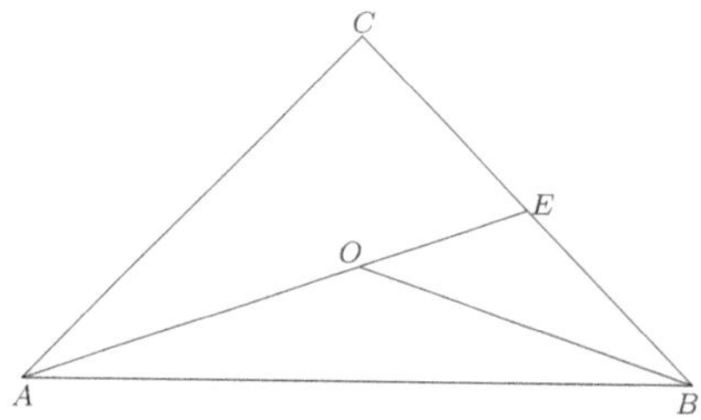

Let CA and CB be two lines drawn from the point C to the extremities of the straight line AB. Let OA and OB be two lines similarly drawn, but included by CA and CB.

To prove that

$$CA + CB > OA + OB.$$

Proof. Produce AO to meet the line CB at E.

Then

$$CA + CE > OA + OE,$$

and

$$BE + OE > OB,$$ § 49

(*a straight line is the shortest line from one point to another*).

Add these inequalities, and we have

$$CA + CE + BE + OE > OA + OE + OB.$$ Ax. 4

Substitute for $CE + BE$ its equal CB, then

$$CA + CB + OE > OA + OE + OB.$$

Take away OE from each side of the inequality.

$$CA + CB > OA + OB.$$ Ax. 5

Q.E.D.

PROPOSITION IX. THEOREM.

101. *Of two straight lines drawn from the same point in a perpendicular to a given line, cutting off on the line unequal segments from the foot of the perpendicular, the more remote is the greater.*

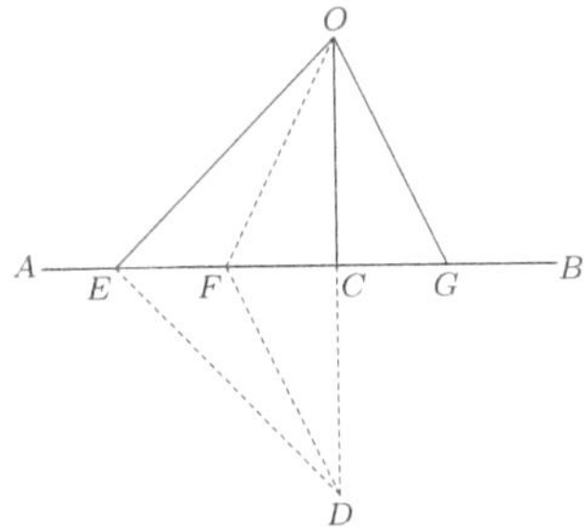

Let OC be perpendicular to AB, OG and OE two straight lines to AB, and CE greater than CG.

To prove that

$$OE > OG.$$

Proof. Take CF equal to CG, and draw OF.

Then

$$OF = OG, \qquad \S\ 95$$

(*two straight lines drawn from a point in a $\perp$ to a line, cutting off on the line equal segments from the foot of the $\perp$, are equal*).

Produce OC to D, making $CD = OC$.

Draw ED and FD.

Then

$$OE = ED, \text{ and } OF = FD. \qquad \S\ 95$$

But

$$OE + ED > OF + FD, \qquad \S\ 100$$

$$\therefore 2OE > 2OF,\ OE > OF, \text{ and } OE > OG. \qquad \text{Q.E.D.}$$

102. COR. *Only two equal straight lines can be drawn from a point to a straight line; and of two unequal lines, the greater cuts off on the line the greater segment from the foot of the perpendicular.*

PARALLEL LINES.

103. Def. Two **parallel lines** are lines that lie in the same plane and cannot meet however far they are produced.

Proposition X. Theorem.

104. *Two straight lines in the same plane perpendicular to the same straight line are parallel.*

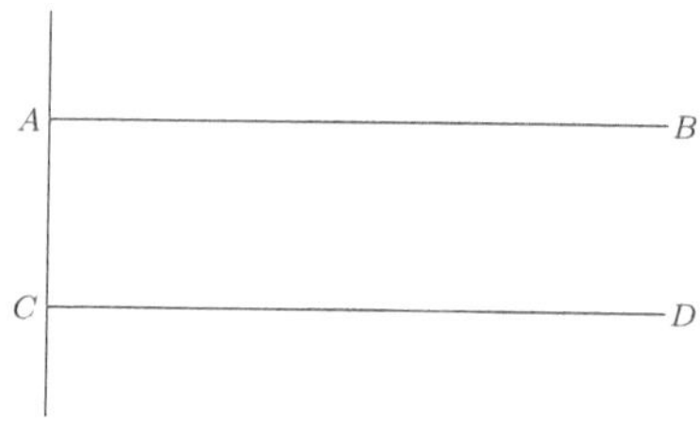

Let AB and CD be perpendicular to AC.

To prove that AB and CD are parallel.

Proof. If AB and CD are not parallel, they will meet if sufficiently prolonged; and we shall have two perpendicular lines from their point of meeting to the same straight line; but this is impossible, § 96

(*only one perpendicular can be drawn to a given line from a given external point*).

$\therefore$ AB and CD are parallel. Q.E.D.

105. Axiom. *Through a given point only one straight line can be drawn parallel to a given straight line.*

106. Cor. *Two straight lines in the same plane parallel to a third straight line are parallel to each other.*

For if they could meet, we should have two straight lines from the point of meeting parallel to a straight line; but this is impossible. § 105

Proposition XI. Theorem.

107. *If a straight line is perpendicular to one of two parallel lines, it is perpendicular to the other also.*

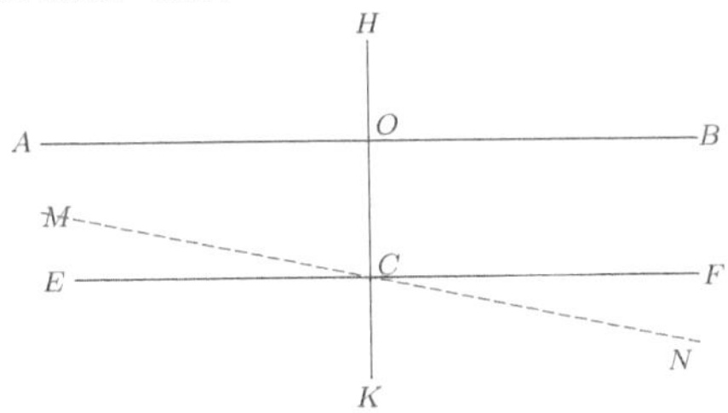

Let AB and EF be two parallel lines, and let HK be perpendicular to AB, and cut EF at C.

To prove that

$$HK \text{ is } \perp \text{ to } EF.$$

Proof. Suppose MN drawn through $C \perp$ to HK.

Then

MN is $\parallel$ to AB. § 104

But

EF is $\parallel$ to AB. Hyp.

$\therefore EF$ coincides with MN. § 105

But

MN is $\perp$ to HK. Const.

$\therefore EF$ is $\perp$ to HK,

that is,

HK is $\perp$ to EF. Q.E.D.

108. Def. A straight line that cuts two or more straight lines is called a **transversal** of those lines.

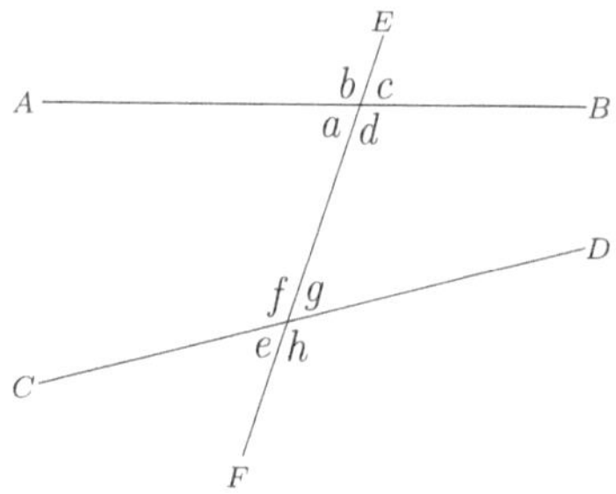

109. If the transversal EF cuts AB and CD, the angles a, d, g, f are called *interior* angles; b, c, h, e are called *exterior* angles.

The angles d and f, and a and g, are called *alternate-interior* angles; the angles b and h, and c and e, are called *alternate-exterior* angles.

The angles b and f, c and g, e and a, h and d, are called *exterior-interior* angles.

Proposition XII. Theorem.

110. *If two parallel lines are cut by a transversal, the alternate-interior angles are equal.*

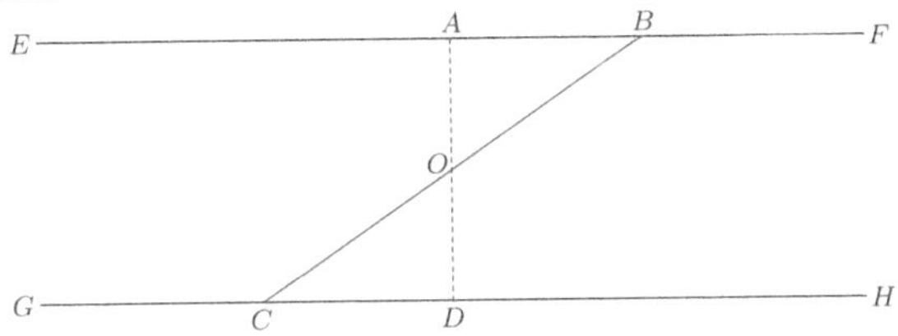

Let EF and GH be two parallel lines cut by the transversal BC.

To prove that

$$\angle EBC = \angle BCH.$$

Proof. Through O, the middle point of BC, suppose AD drawn $\perp$ to GH.

Then

AD is likewise $\perp$ to EF, § 107

(*a straight line* $\perp$ *to one of two* $\|_s$ *is* $\perp$ *to the other*),

that is,

CD and BA are both $\perp$ to AD.

Apply the figure COD to the figure BOA, so that OD shall fall along OA.

Then

OC will fall along OB, § 93

(*since* $\angle COD = \angle BOA$, *being vertical* $\angle_s$);

and

C will fall on B,

(*since* $OC = OB$, *by construction*).

Then

the $\perp$ CD will fall along the $\perp$ BA, § 96

(*only one* $\perp$ *can be drawn to a given line from a given external point*).

$\therefore \angle OCD$ coincides with $\angle OBA$, and is equal to it, § 60

(*two angles are equal, if their vertices coincide and their sides coincide, each with each*).

Q.E.D.

Proposition XIII. Theorem.

111. Conversely: *When two straight lines in the same plane are cut by a transversal, if the alternate-interior angles are equal, the two straight lines are parallel.*

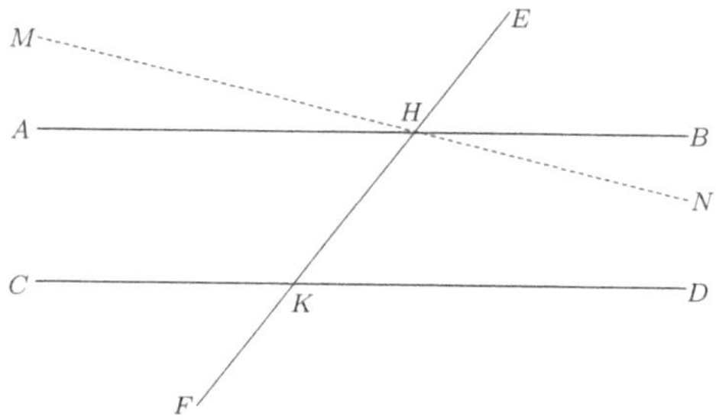

Let EF cut the straight lines AB and CD in the points H and K, and let the angles AHK and HKD be equal.

To prove that

AB is $\parallel$ to CD.

Proof. Suppose MN drawn through H $\parallel$ to CD.

Then

$\angle MHK = \angle HKD$, § 110

(being alt.-int. $\angle_s$ *of* $\parallel$ *lines).*

But

$\angle AHK = \angle HKD$. Hyp.

$\therefore \angle MHK = \angle AHK$. Ax. 1

$\therefore MN$ and AB coincide. § 60

But

MN is $\parallel$ to CD. Const.

$\therefore AB$, which coincides with MN, is $\parallel$ to CD. Q.E.D.

Ex. 5. Find the complement and the supplement of an angle that contains $37° \, 53' \, 49''$.

Ex. 6. If the complement of an angle is one third of its supplement, how many degrees does the angle contain?

Proposition XIV. Theorem.

112. *If two parallel lines are cut by a transversal, the exterior-interior angles are equal.*

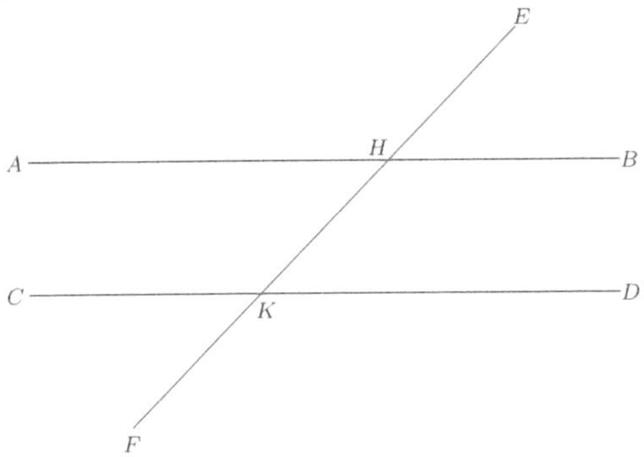

Let AB and CD be two parallel lines cut by the transversal EF, in the points H and K.

To prove that

$$\angle EHB = \angle HKD.$$

Proof.

$$\angle EHB = \angle AHK, \quad \text{§ 93}$$

(*being vertical* $\angle_s$).

$$\angle AHK = \angle HKD, \quad \text{§ 110}$$

(*being alt.-int.* $\angle_s$ *of* $\|$ *lines*).

$$\therefore \angle EHB = \angle HKD. \quad \text{Ax. 1}$$

In like manner

$$\angle EHA = \angle HKC. \quad \text{Q.E.D.}$$

113. Cor. *The alternate-exterior angles EHB and CKF, and also AHE and DKF, are equal.*

Proposition XV. Theorem.

114. Conversely: *When two straight lines in a plane are cut by a transversal, if the exterior-interior angles are equal, these two straight lines are parallel.*

(Proof similar to that in § 111.)

Proposition XVI. Theorem.

115. *If two parallel lines are cut by a transversal, the two interior angles on the same side of the transversal are supplementary.*

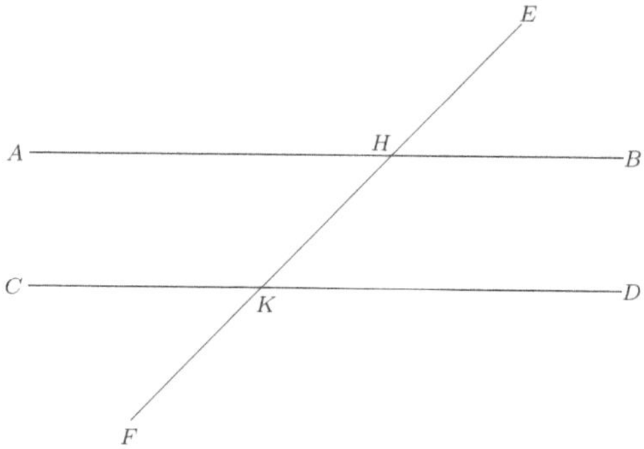

Let AB and CD be two parallel lines cut by the transversal EF in the points H and K.

To prove that $\angle_s BHK$ *and* HKD *are supplementary.*

Proof.

$$\angle EHB + \angle BHK = \text{a st. } \angle, \qquad \text{§ 86}$$

(being sup.-adj. $\angle_s$*).*

But

$$\angle EHB = \angle HKD, \qquad \text{§ 112}$$

(being ext.-int. $\angle_s$ *of* $\|$ *lines).*

$$\therefore \angle BHK + \angle HKD = \text{a st. } \angle.$$

$$\therefore \angle_s BHK \text{ and } HKD \text{ are supplementary.} \qquad \text{§ 77}$$

Q.E.D.

PROPOSITION XVII. THEOREM.

116. CONVERSELY: *When two straight lines in a plane are cut by a transversal, if two interior angles on the same side of the transversal are supplementary, the two straight lines are parallel.*

(Proof similar to that in § 111.)

TRIANGLES.

117. A **triangle** is a portion of a plane bounded by three straight lines; as, ABC (Fig. 1).

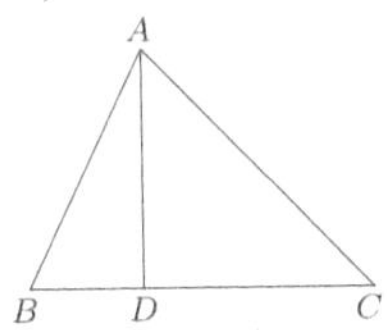

FIG. 1.

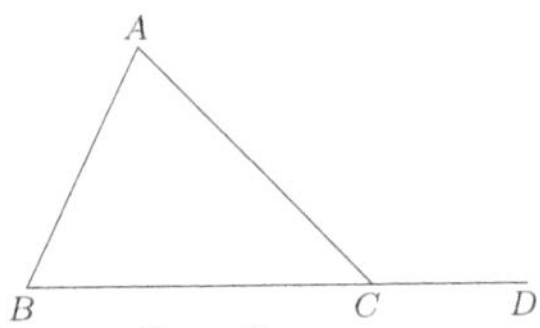

FIG. 2.

The bounding lines are called the **sides** of the triangle, and their sum is called its **perimeter**; the angles included by the sides are called the **angles** of the triangle, and the vertices of these angles, the **vertices** of the triangle.

118. Adjacent angles of a rectilinear figure are two angles that have one side of the figure common; as, angles A and B (Fig. 2).

119. An **exterior angle** of a triangle is an angle included by one side and another side produced; as, ACD (Fig. 2). The interior angle ACB is adjacent to the exterior angle; the interior angles, A and B, are called **opposite interior angles**.

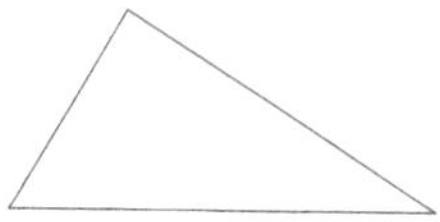

Scalene.

Isosceles.

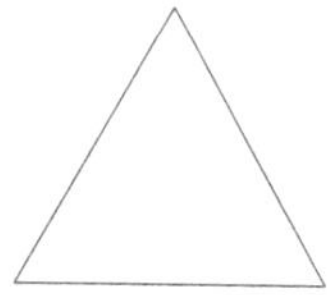

Equilateral.

120. A triangle is called a **scalene triangle** when no two of its sides are equal; an **isosceles triangle**, when two of its sides are equal; an **equilateral triangle**, when its three sides are equal.

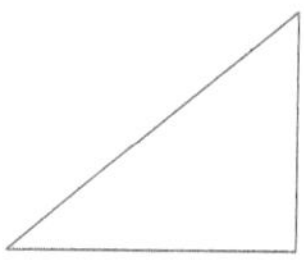

Right.

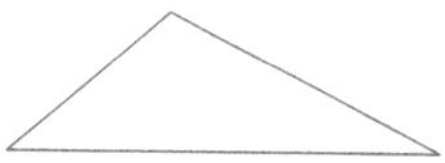

Obtuse.

Acute.

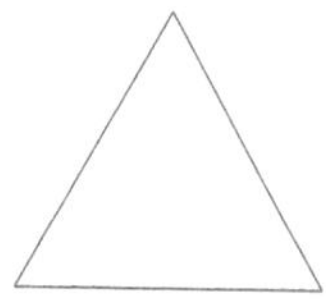

Equiangular.

121. A triangle is called a **right triangle**, when one of its angles is a right angle; an **obtuse triangle**, when one of its angles is an obtuse angle; an **acute triangle**, when all three of its angles are acute angles; an **equiangular triangle**, when its three angles are equal.

122. In a right triangle, the side opposite the right angle is called the **hypotenuse**, and the other two sides the **legs**.

123. The side on which a triangle is supposed to stand is called the base of the triangle. In the isosceles triangle, the equal sides are called the legs, and the other side, the base; in other triangles, any one of the sides may be taken as the base.

124. The angle opposite the base of a triangle is called the **vertical angle**, and its vertex, the **vertex** of the triangle.

125. The **altitude** of a triangle is the perpendicular from the vertex to the base, or to the base produced; as, AD (Fig. 1).

126. The three perpendiculars from the vertices of a triangle to the opposite sides (produced if necessary) are called the **altitudes** of the triangle; the three bisectors of the angles are called the **bisectors** of the triangle; and the three lines from the vertices to the middle points of the opposite sides are called the **medians** of the triangle.

127. If two triangles have the angles of the one equal, respectively, to the angles of the other, the equal angles are called **homologous angles**, and the sides opposite the equal angles are called **homologous sides**.

128. Two triangles are equal in all respects if they can be made to coincide (§ 60). The homologous sides of *equal triangles* are equal, and the homologous angles are equal.

PROPOSITION XVIII. THEOREM.

129. *The sum of the three angles of a triangle is equal to two right angles.*

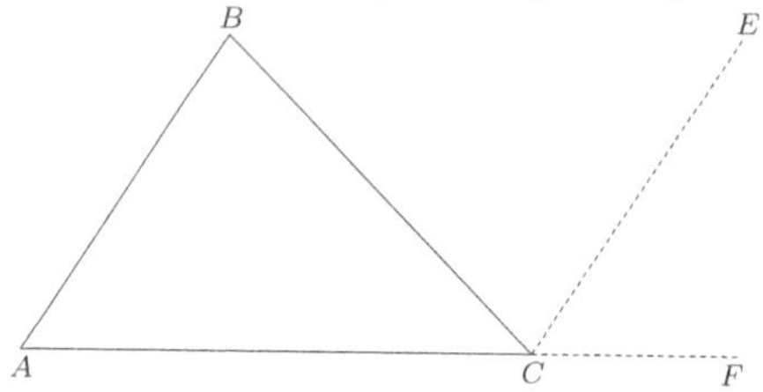

Let A, B, and BCA be the angles of the triangle ABC.

To prove that $\angle A + \angle B + \angle BCA = 2$ *rt.* $\angle_s$.

Proof. Suppose CE drawn $\parallel$ to AS, and prolong AC to F.

Then

$$\angle ECF + \angle ECB + \angle BCA = 2 \text{ rt. } \angle_s, \qquad \S\ 89$$

(*the sum of all the* $\angle_s$ *about a point on the same side of a straight line passing through the point is equal to* 2 *rt.* $\angle_s$).

But

$$\angle A = \angle ECF, \qquad \S\ 112$$

(*being ext.-int.* $\angle_s$ *of the* $\parallel$ *lines* AB *and* CE),

and

$$\angle B = \angle BCE, \qquad \S\ 110$$

(*being alt.-int.* $\angle_s$ *of the* $\parallel$ *lines* AB *and* CE).

Put for the $\angle_s ECF$ and BCE their equals, the $\angle_s A$ and B.

Then

$$\angle A + \angle B + \angle BCA = 2 \text{ rt. } \angle_s. \qquad \text{Q.E.D.}$$

130. COR. 1. *The sum of two angles of a triangle is less than two right angles.*

131. COR. 2. *If the sum of two angles of a triangle is taken from two right angles, the remainder is equal to the third angle.*

132. COR. 3. *If two triangles have two angles of the one equal to two angles of the other, the third angles are equal.*

133. COR. 4. *If two right triangles have an acute angle of the one equal to an acute angle of the other, the other acute angles are equal.*

134. Cor. 5. *In a triangle there can be but one right angle, or one obtuse angle.*

135. Cor. 6. *In a right triangle the two acute angles are together equal to one right angle, or* $90°$.

136. Cor. 7. *In an equiangular triangle, each angle is one third of two right angles, or* $60°$.

137. Cor. 8. *An exterior angle of a triangle is equal to the sum of the two opposite interior angles, and therefore greater than either of them.*

Proposition XIX. Theorem.

138. *The sum of two sides of a triangle is greater than the third side, and their difference is less than the third side.*

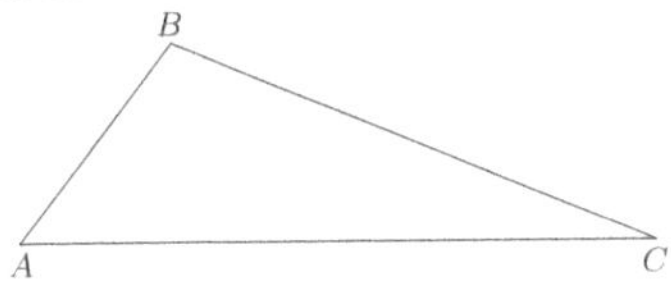

In the triangle ABC, let AC be the longest side.

To prove that $AB + BC > AC$, *and* $AC - BC < AB$.

Proof.

$$AB + BC > AC, \qquad \text{§ 49}$$

(*a straight line is the shortest line from one point to another*).

Take away BC from both sides.

Then

$$AB > AC - BC, \qquad \text{Ax. 5}$$

or

$$AC - BC < AB. \qquad \text{Q.E.D.}$$

PROPOSITION XX. THEOREM.

139. *Two triangles are equal if two angles and the included side of the one are equal, respectively, to two angles and the included side of the other.*

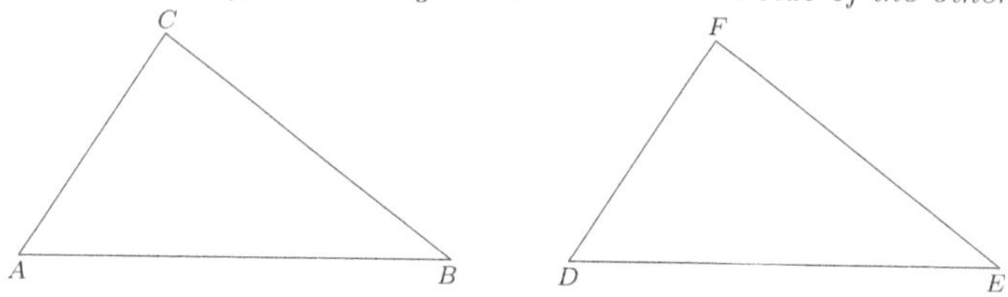

In the triangles ABC, DEF, let the angle A be equal to the angle D, B to E, and the side AB to DE.

To prove that

$$\triangle ABC = \triangle DEF.$$

Proof. Apply the $\triangle ABC$ to the $\triangle DEF$ so that AB shall coincide with its equal, DE.

Then

AC will fall along DF, and BC along EF,
(*for* $\angle A = \angle D$, *and* $\angle B = \angle E$, *by hyp.*).

$\therefore C$ will fall on F, § 48
(*two straight lines can intersect in only one point*).

$\therefore$ the two $\triangle_s$ coincide, and are equal. § 60

Q.E.D.

140. COR. 1. *Two triangles are equal if a side and any two angles of the one are equal to the homologous side and two angles of the other.* § 132

141. COR. 2. *Two right triangles are equal if the hypotenuse and an acute angle of the one are equal, respectively, to the hypotenuse and an acute angle of the other.* § 133

142. COR. 3. *Two right triangles are equal if a leg and an acute angle of the one are equal, respectively, to a leg and the homologous acute angle of the other.* § 133

PROPOSITION XXI. THEOREM.

143. *Two triangles are equal if two sides and the included angle of the one are equal, respectively, to two sides and the included angle of the other.*

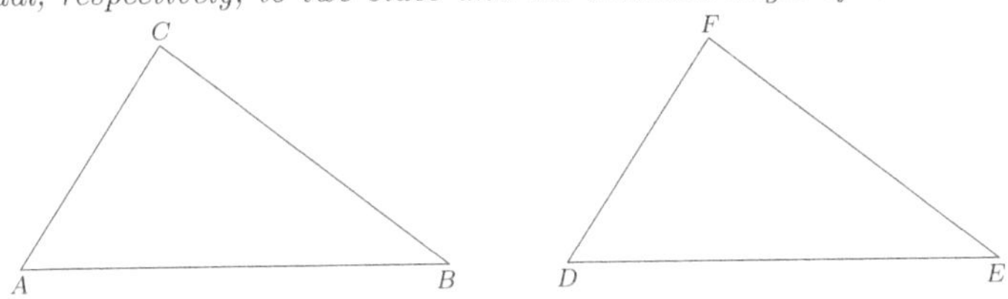

In the triangles ABC and DEF, let AB be equal to DE, AC to DF, and the angle A to the angle D.

To prove that

$$\triangle ABC = \triangle DEF.$$

Proof. Apply the $\triangle ABC$ to the $\triangle DEF$ so that AB shall coincide with its equal, DE.

Then AC will fall along DF,
(for $\angle A = \angle D$, by hyp.);

and C will fall on F,
(for $AC = DF$, by hyp.).

$\therefore CB = FE$,
(their extremities being the same points).

$\therefore$ the two $\triangle_s$ coincide, and are equal. Q.E.D.

144. COR. *Two right triangles are equal if their legs are equal, each to each.*

NOTE. In § 139 we have given two angles and the included side, in § 143 two sides and the included angle; hence, by interchanging the words *sides* and *angles*, either theorem is changed to the other. This is called the *Principle of Duality*, or the *Principle of Reciprocity*. The reciprocal of a theorem is not always true, just as the converse of a theorem is not always true.

PROPOSITION XXII. THEOREM.

145. *In an isosceles triangle the angles opposite the equal sides are equal.*

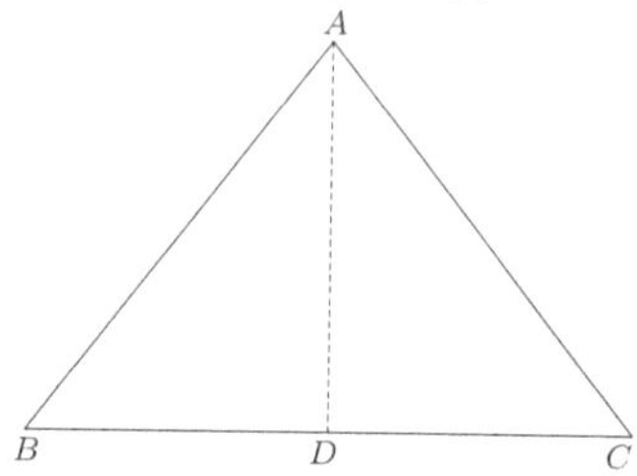

Let ABC be an isosceles triangle, having AB and AC equal.
To prove that

$$\angle B = \angle C.$$

Proof. Suppose AD drawn so as to bisect the $\angle BAC$.

In the $\triangle_s ADB$ and ADC,

$$AB = AC, \qquad \text{Hyp.}$$

$$AD = AD, \qquad \text{Iden.}$$

$$\text{and } \angle BAD = \angle CAD. \qquad \text{Const.}$$

$$\therefore \triangle ADB = \triangle ADC, \qquad \S\ 143$$

(*two $\triangle_s$ are equal if two sides and the included $\angle$ of the one are equal, respectively, to two sides and the included $\angle$ of the other*).

$$\therefore \angle B = \angle C, \qquad \S\ 128$$

(*being homologous angles of equal triangles*).

Q.E.D.

146. COR. *An equilateral triangle is equiangular, and each angle is two thirds of a right angle.*

Ex. 7. If the equal sides of an isosceles triangle are produced, the angles on the other side of the base are equal.

PROPOSITION XXIII. THEOREM.

147. *If two angles of a triangle are equal, the sides opposite the equal angles are equal, and the triangle is isosceles.*

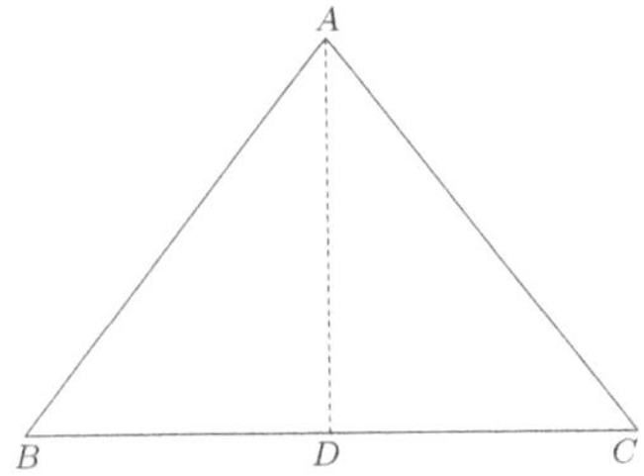

In the triangle ABC, let the angle B be equal to the angle C.

To prove that

$$AB = AC.$$

Proof.

Suppose AD drawn $\perp$ to BC.

In the rt. $\triangle_s ADB$ and ADC,

$$AD = AD, \qquad \text{Iden.}$$

$$\text{and } \angle B = \angle C. \qquad \text{Hyp.}$$

$$\therefore \text{rt. } \triangle ADB = \text{rt. } \triangle ADC, \qquad \S\ 142$$

(*having a leg and an acute $\angle$ of the one equal, respectively, to a leg and the homologous acute $\angle$ of the other*).

$$\therefore AB = AC, \qquad \S\ 128$$

(*being homologous sides of equal $\triangle_s$*).

Q.E.D.

148. COR. 1. *An equiangular triangle is also equilateral.*

149. COR. 2. *The perpendicular from the vertex to the base of an isosceles triangle bisects the base, and bisects the vertical angle of the triangle.*

Proposition XXIV. Theorem.

150. *Two triangles are equal if the three sides of the one are equal, respectively, to the three sides of the other.*

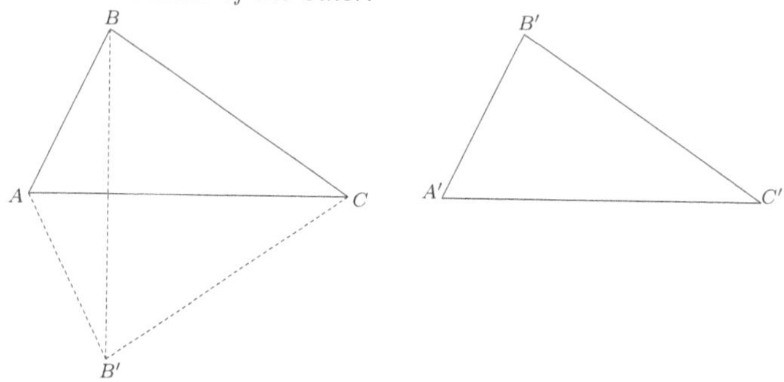

In the triangles ABC and $A'B'C'$, let AB be equal to $A'B'$, AC to $A'C'$, BC to $B'C'$.

To prove that

$$\triangle ABC = \triangle A'B'C'.$$

Proof. Place $\triangle A'B'C'$ in the position $\triangle AB'C$ having its greatest side $\triangle A'C'$ in coincidence with its equal $\triangle AC$, and its vertex at B', opposite B; and draw BB'.

Since

$$AB = AB' \qquad \text{Hyp.}$$

$$\angle ABB' = \angle AB'B \qquad \S\ 145$$

(*in an isosceles $\triangle$ the $\angle_s$ opposite the equal sides are equal*).

Since

$$CB = CB', \qquad \text{Hyp.}$$

$$\angle CBB' = \angle CB'B. \qquad \S\ 145$$

$$\therefore \angle ABB' + \angle CBB' = \angle AB'B + \angle CB'B. \qquad \text{Ax. 2}$$

Hence,

$$\angle ABC = \angle AB'C.$$

$$\therefore \triangle ABC = \triangle AB'C, \qquad \S\ 143$$

(*two $\triangle_s$ are equal if two sides and the included $\angle$ of the one are equal, respectively, to two sides and the included $\angle$ of the other*).

$$\therefore \triangle ABC = \triangle A'B'C'. \qquad \text{Q.E.D.}$$

Proposition XXV. Theorem.

151. *Two right triangles are equal if a leg and the hypotenuse of the one are equal, respectively, to a leg and the hypotenuse of the other.*

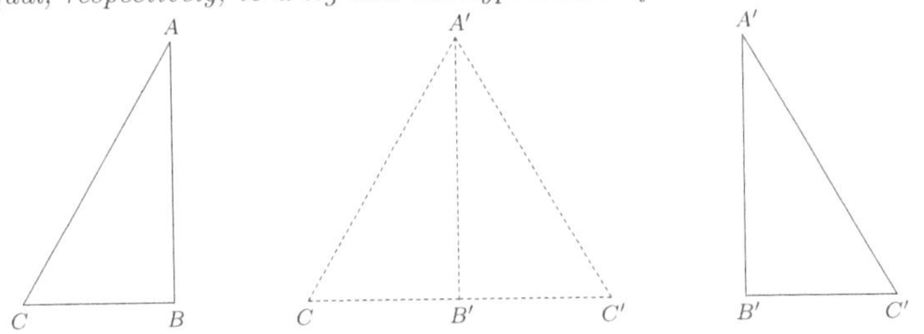

In the right triangles ABC and $A'B'C'$, let AB be equal to $A'B'$, and AC to $A'C'$.

To prove that

$$\triangle ABC = \triangle A'B'C'.$$

Proof. Apply the $\triangle ABC$ to the $\triangle A'B'C'$, so that AB shall coincide with $A'B'$, A falling on A', B on B', and C and C' on opposite sides of $A'B'$.

Then

BC will fall along $C'B'$ produced,

(*for* $\angle ABC = \angle A'B'C'$, *each being a rt.* $\angle$.).

Since

$AC = A'C'$, Hyp.

the $\triangle A'CC'$ is an isosceles triangle. § 120

$\therefore \angle C = \angle C'$, § 145

($\angle_s$ *opposite the equal sides of an isosceles* $\triangle$ *are equal*).

$\therefore \triangle_s ABC$ and $A'B'C'$ are equal, § 141

(*two right* $\triangle_s$ *are equal if they have the hypotenuse and an acute* $\angle$ *of, the one equal to the hypotenuse and an acute* $\angle$ *of the other*).

Q.E.D.

Ex. 8. How many degrees are there in each of the acute angles of an isosceles right triangle?

Proposition XXVI. Theorem.

152. *If two sides of a triangle are unequal, the angles opposite are unequal, and the greater angle is opposite the greater side.*

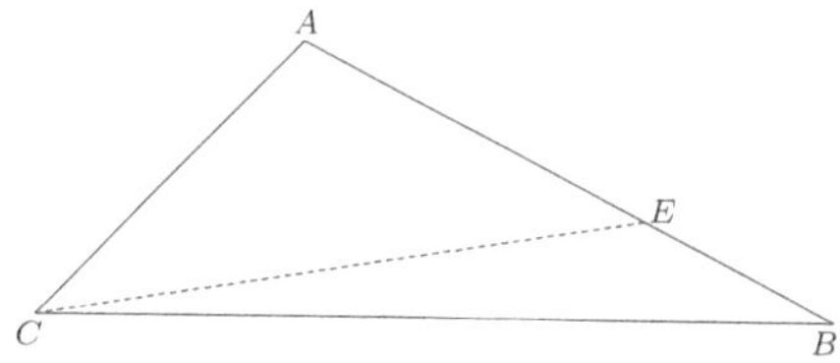

In the triangle ACB, let AB be greater than AC.

To prove that $\angle ACB$ is greater than $\angle B$.

Proof.

On AB take AE equal to AC.

Draw EC.

$\angle AEC = \angle ACE$ § 145

(*being* $\angle_s$ *opposite equal sides*).

But

$\angle AEC$ is greater than $\angle B$ § 137

(*an exterior* $\angle$ *of a* $\triangle$ *is greater than either opposite interior* $\angle$),

and

$\angle ACB$ is greater than $\angle ACE$. Ax. 8

Substitute for $\angle ACE$ its equal $\angle AEC$,

then

$\angle ACB$ is greater than $\angle AEC$.

Since $\angle AEC$ is greater than $\angle B$, and $\angle ACB$ is greater than $\angle AEC$,

$\angle ACB$ is greater than $\angle B$. Q.E.D.

Ex. 9. If any angle of an isosceles triangle is equal to two thirds of a right angle (60°), what is the value of each of the two remaining angles?

Ex. 10. One angle of a triangle is 34°. Find the other angles, if one of them is twice the other.

PROPOSITION XXVII. THEOREM.

153. RECIPROCALLY: *If two angles of a triangle are unequal, the sides opposite are unequal, and the greater side is opposite the greater angle.*

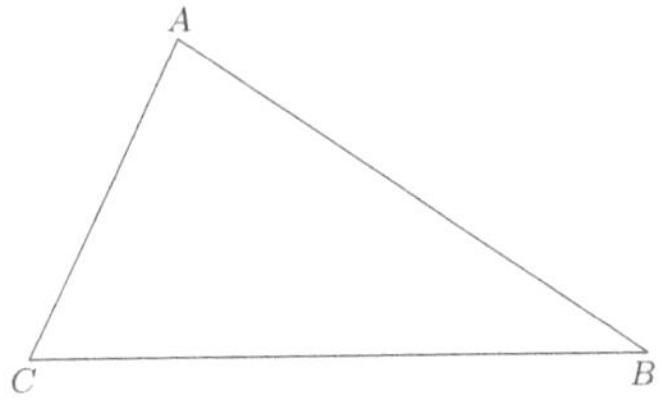

In the triangle ACB, let the angle C be greater than the angle B.

To prove that $AB > AC$.

Proof.

Now $AB = AC$, or $< AC$, or $> AC$.

But AB is not equal to AC;

for then the $\angle C$ would be equal to the $\angle B$, § 145
(*being $\angle_s$ opposite equal sides*).

And AB is not less than AC;

for then the $\angle C$ would be less than the $\angle B$. § 152

Both these conclusions are contrary to the hypothesis that the $\angle C$ is greater than the $\angle B$.

Hence, AB cannot be equal to AC or less than AC.

$\therefore AB > AC$. Q.E.D.

Ex. 11. If the vertical angle of an isosceles triangle is equal to 30°, find the exterior angle included by a side and the base produced.

Ex. 12. If the vertical angle of an isosceles triangle is equal to 36°, find the angle included by the bisectors of the base angles.

PROPOSITION XXVIII. THEOREM.

154. *If two triangles have two sides of the one equal, respectively, to two sides of the other, but the included angle of the first triangle greater than the included angle of the second, then the third side of the first is greater than the third side of the second.*

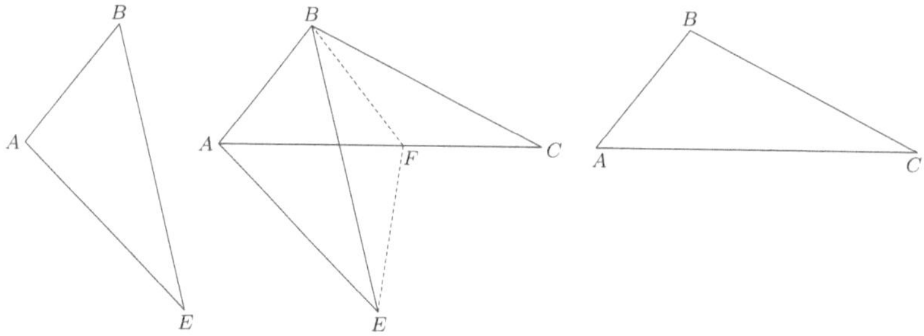

In the triangles ABC and ABE, let AB be equal to AB, BC to BE; but let the angle ABC be greater than the angle ABE.

To prove that

$$AC > AE.$$

Proof. Place the $\triangle_s$ so that AB of the one shall fall on AB of the other, and BE within the $\angle ABC$.

Suppose BF drawn to bisect the $\angle EBC$, and draw EF.

The $\triangle_s EBF$ and CBF are equal. § 143

For

$$BF = BF, \qquad \text{Iden.}$$

$$BE = BC, \qquad \text{Hyp.}$$

and

$$\angle EBF = \angle CBF. \qquad \text{Const.}$$

$$\therefore EF = FC. \qquad \text{§ 128}$$

Now

$$AF + FE > AE. \qquad \text{§ 138}$$

$$\therefore AF + FC > AE.$$

$$\therefore AC > AE. \qquad \text{Q.E.D.}$$

PROPOSITION XXIX. THEOREM.

155. CONVERSELY*: If two sides of a triangle are equal, respectively, to two sides of another, but the third side of the first triangle is greater than the third side of the second, then the angle opposite the third side of the first triangle is greater than the angle opposite the third side of the second.*

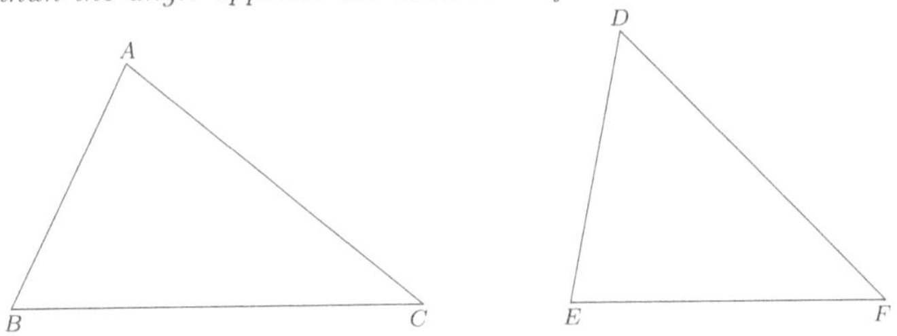

In the triangles ABC and DEF, let AB be equal to DE, AC to DF; but let BC be greater than EF.

To prove that the $\angle A$ is greater than the $\angle D$.

Proof. Now the $\angle A$ is equal to the $\angle D$, or less than the $\angle D$, or greater than the $\angle D$.

But the $\angle A$ is not equal to the $\angle D$;

for then the $\triangle ABC$ would be equal to the $\triangle DEF$, § 143

(having two sides and the included $\angle$ of the one equal, respectively, to two sides and the included $\angle$ of the other),

and BC would be equal to EF.

And the $\angle A$ is not less than the $\angle D$, for then BC would be less than EF. § 154

Both these conclusions are contrary to the hypothesis that BC is greater than EF.

Since the $\angle A$ is not equal to the $\angle D$ or less than the $\angle D$,

the $\angle A$ is greater than the $\angle D$. Q.E.D.

LOCI OF POINTS.

156. If it is required to find a point which shall fulfil a *single* geometric condition, the point may have an *unlimited number of positions.* If, however, all the points are in the same plane, the required point will be confined to a *particular line*, or *group of lines.*

A point in a plane at a given distance from a fixed straight line of indefinite length in that plane, is evidently in one of two straight lines, so drawn as to be everywhere at the given distance from the fixed line, one on one side of the fixed line, and the other on the other side of it.

A point in a plane equidistant from two parallel lines in that plane is evidently in a straight line drawn between the two given parallel lines and everywhere equidistant from them.

157. All points in a plane that satisfy a single geometrical condition lie, in general, in a line or group of lines; and this line or group of lines is called the **locus** of the points that satisfy the given condition.

158. To prove *completely* that a certain line is the locus of points that fulfil a given condition, it is necessary to prove

1. *Any point in the line satisfies the given condition; and any point not in the line does not satisfy the given condition.*

Or, to prove

2. *Any point that satisfies the given condition lies in the line; and any point in the line satisfies the given condition.*

NOTE. The word *locus* (pronounced lo´kus) is a Latin word that signifies *place.* The plural of locus is loci (pronounced lo´si).

159. DEF. A line which bisects a given line and is perpendicular to it is called the **perpendicular bisector** of the line.

PROPOSITION XXX. THEOREM.

160. *The perpendicular bisector of a given line is the locus of points equidistant from the extremities of the line.*

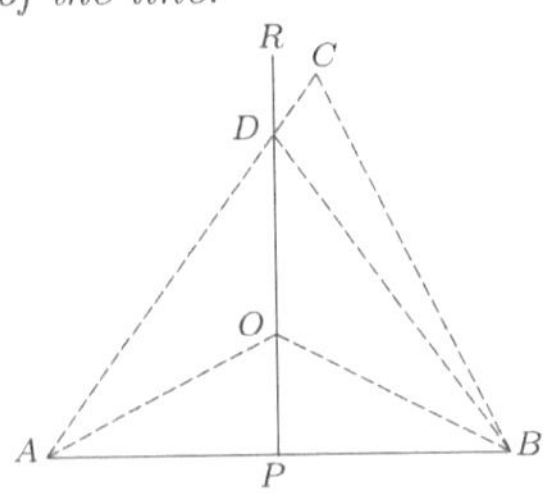

Let PR be the perpendicular bisector of the line AB, O any point in PR, and C any point not in PR.

Draw OA and OB, CA and CB.

To prove OA and OB equal, CA and CB unequal.

Proof.

1. $\triangle OPA = \triangle OPB$, § 144

for $PA = PB$ by hypothesis, and OP is common,
(*two right $\triangle_s$ are equal if their legs are equal, each to each*).

$$\therefore OA = OB.$$ § 128

2. Since C is not in the $\perp$, CA or CB will cut the $\perp$.

Let CA cut the $\perp$ at D, and draw DB.

Then, by the first part of the proof $DA = DB$.

But

$$CB < CD + DB.$$ § 138

$$\therefore CB < CD + DA.$$

That is,

$$CB < CA.$$

$\therefore PR$ is the locus of points equidistant from A and B. § 158,1

Q.E.D.

161. Cor. *Two points each equidistant from the extremities of a line determine the perpendicular bisector of the line.*

Proposition XXXI. Theorem.

162. *The bisector of a given angle is the locus of points equidistant from the sides of the angle.*

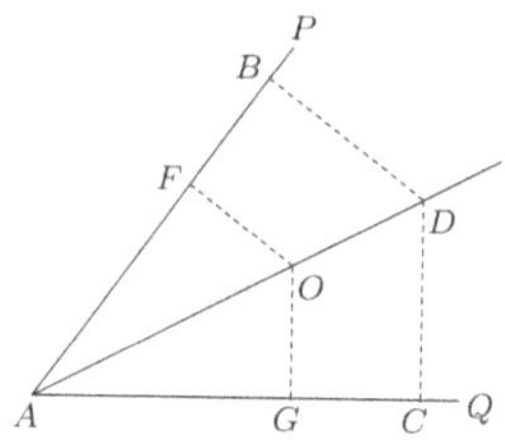

Let O be any point equidistant from the sides of the angle PAQ.

To prove that O is in the bisector of the $\angle PAQ$.

Proof.

Draw AO.

Suppose OF drawn $\perp$ to AP and $OG \perp$ to AQ.

In the rt. $\triangle_s AFO$ and AGO,

$AO = AO$, Iden.

$OF = OG$, Hyp.

$\therefore \triangle AFO = \triangle AGO$. § 151

$\therefore \angle FAO = \angle GAO$. § 128

$\therefore O$ is in the bisector of the $\angle PAQ$.

Let D be any point in the bisector of the angle PAQ.

To prove that D is equidistant from AP and AQ.

Proof.

Suppose DB drawn $\perp$ to AP and $DC \perp$ to AQ.

In the rt. $\triangle_s ABD$ and ACD,

$AD = AD$, Iden.

$\angle DAB = \angle DAC$, Hyp.

$\therefore \triangle ABD = \triangle ACD$. § 141

$\therefore DB = DC$. § 128

$\therefore D$ is equidistant from AP and AQ.

∴ the bisector of the $\angle PAQ$ is the locus of points that are equidistant from its sides. § 158, 2

QUADRILATERALS.

163. A **quadrilateral** is a portion of a plane bounded by four straight lines. The bounding lines are the **sides**, the angles formed by these sides are the **angles**, and the vertices of these angles are the **vertices**, of the quadrilateral.

164. A **trapezium** is a quadrilateral which has no two sides parallel.

165. A **trapezoid** is a quadrilateral which has two sides, and only two sides, parallel.

166. A **parallelogram** is a quadrilateral which has its opposite sides parallel.

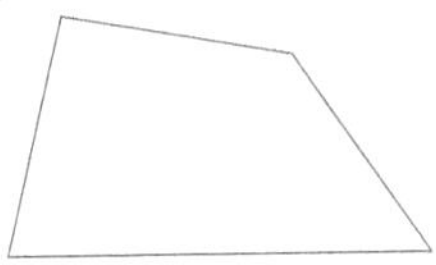

Trapezium.

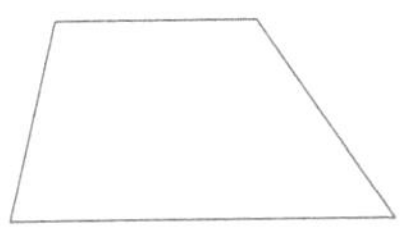

Trapezoid.

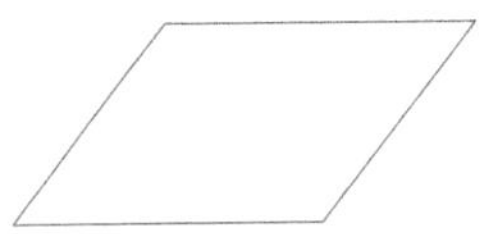

Parallelogram.

167. A **rectangle** is a parallelogram which has its angles right angles.

168. A **square** is a rectangle which has its sides equal.

169. A **rhomboid** is a parallelogram which has its angles oblique angles.

170. A **rhombus** is a rhomboid which has its sides equal.

Square.

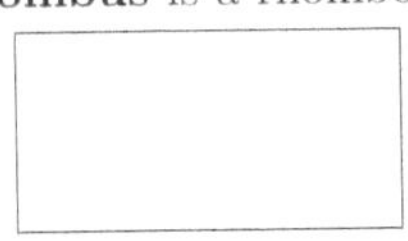

Rectangle.

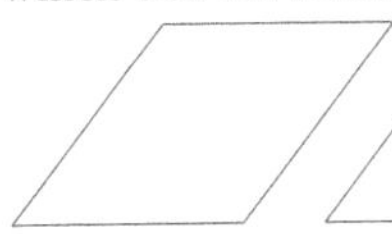

Rhombus.

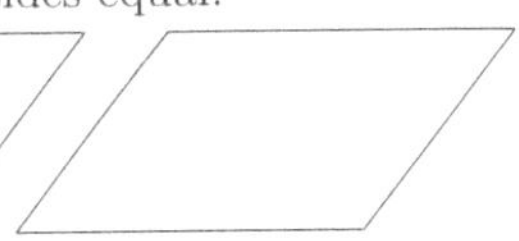

Rhomboid.

171. The side upon which a parallelogram stands, and the opposite side, are called its lower and upper *bases*.

172. Two parallel sides of a trapezoid are called its **bases**, the other two sides its **legs**, and the line joining the middle points of the legs is called the **median** of the trapezoid.

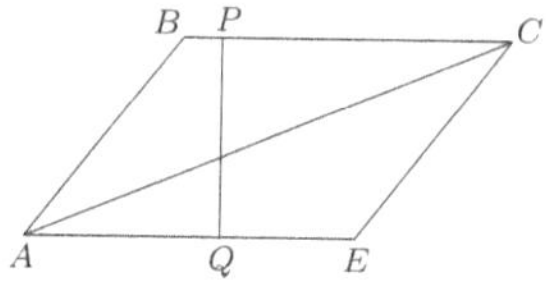

173. A trapezoid is called an **isosceles trapezoid** if its legs are equal.

174. The **altitude** of a parallelogram or trapezoid is the perpendicular distance between its bases, as PQ.

175. A **diagonal** of a quadrilateral is a straight line joining two opposite vertices, as AC.

PROPOSITION XXXII. THEOREM.

176. *Two angles whose sides are parallel, each to each, are either equal or supplementary.*

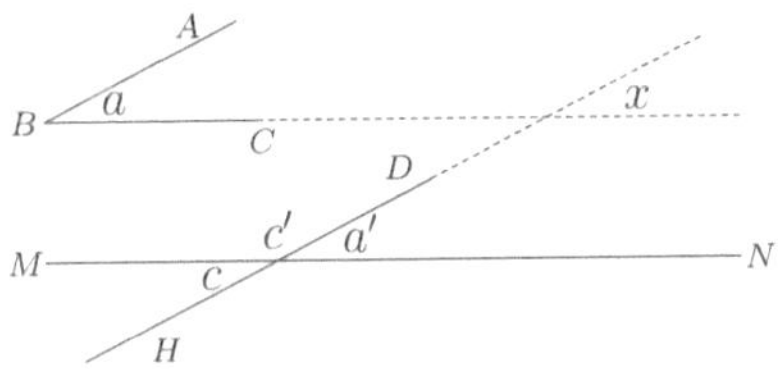

Let BA be parallel to HD, and BC be parallel to MN.

To prove $\angle_s a$, a' *and* c *equal;* a *and* c' *supplementary.*

Proof.

Let HD and BC prolonged intersect at x.

Then $\angle a = \angle x$, and $\angle a' = \angle x$. § 112

$\therefore \angle a = \angle a'$. Ax. 1

Also $\angle c = \angle a'$ (§ 93). $\therefore \angle c = \angle a$. Ax. 1

Now $\angle a'$ and $\angle c'$ are supplementary. § 89

Put $\angle a$ for its equal, $\angle a'$.

Then $\angle a$ and $\angle c'$ are supplementary. Q.E.D.

177. COR. *The opposite angles of a parallelogram are equal, and the adjacent angles are supplementary.*

PROPOSITION XXXIII. THEOREM.

178. *The opposite sides of a parallelogram are equal.*

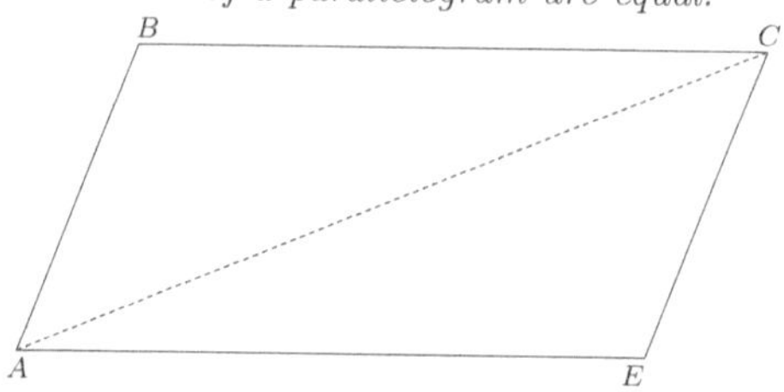

Let the figure $ABCE$ be a parallelogram.

To prove $BC = AE$, and $AB = EC$.

Proof.

Draw the diagonal AC.

$\triangle ABC = \triangle CEA$. § 139

For AC is common,

$\angle BAC = \angle ACE$, and $\angle ACB = \angle CAE$, § 110
(*being alt-int.* $\angle_s$ *of* $\|$ *lines*).

$\therefore BC = AE$, and $AB = CE$, § 128
(*being homologous sides of equal* $\triangle_s$).

Q.E.D.

179. COR. 1. *A diagonal divides a parallelogram into two equal triangles.*

180. COR. 2. *Parallel lines comprehended between parallel lines are equal.*

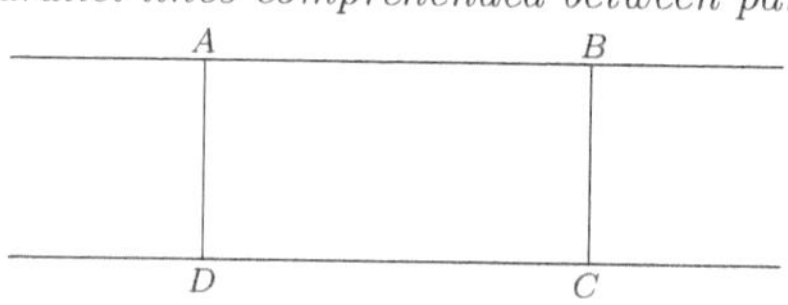

181. COR. 3. *Two parallel lines are everywhere equally distant.*

For if AB and DC are parallel, $\perp_s$ dropped from *any* points in AB to DC, are equal, § 180. Hence, *all* points in AB are equidistant from DC.

Proposition XXXIV. Theorem.

182. *If the opposite sides of a quadrilateral are equal, the figure is a parallelogram.*

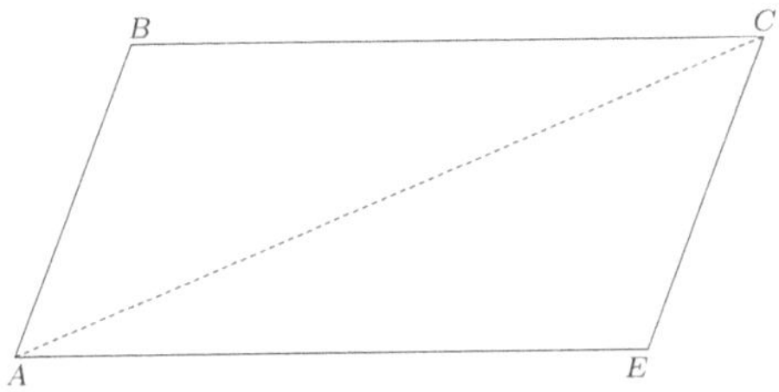

Let the figure $ABCE$ be a quadrilateral, having BC equal to AE and AB to EC.

To prove that the figure $ABCE$ is a ▱.

Proof.

Draw the diagonal AC.

In the $\triangle_s ABC$ and CEA,

$BC = AE$, Hyp.

$AB = CE$, Hyp.

and

$AC = AC$, Iden.

$\therefore \triangle ABC = \triangle CEA$, § 150

(*having three sides of the one equal, respectively, to the three sides of the other*).

$\therefore \angle ACB = \angle CAE$, § 128

and

$\angle BAC = \angle ACE$,

(*being homologous* $\angle_s$ *of equal* $\triangle_s$).

$\therefore BC$ is $\parallel$ to AE,

and

AB is $\parallel$ to EC, § 111

(*two lines in the same plane cut by a transversal are parallel, if the alt.-int.* $\angle_s$ *are equal*).

$\therefore$ the figure $ACBE$ is a ▱, § 166

(*having its opposite sides parallel*). Q.E.D.

Proposition XXXV. Theorem.

183. *If two sides of a quadrilateral are equal and parallel, then the other two sides are equal and parallel, and the figure is a parallelogram.*

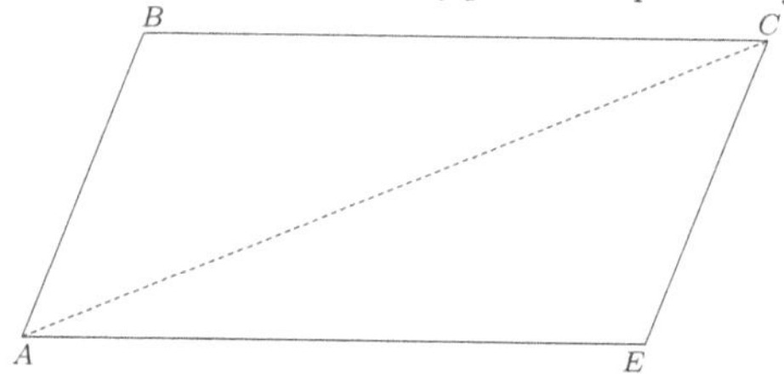

Let the figure $ABCE$ be a quadrilateral, having the side AE equal and parallel to BC.

To prove that AB is equal and parallel to EC.

Proof.

Draw AC.

The $\triangle_s ABC$ and CEA are equal, § 143

(having two sides and the included $\angle$ of each equal, respectively).

For

AC is common,

$BC = AE$ Hyp.

and

$\angle BCA = \angle CAE$, § 110

(being alt.-int. $\angle_s$ of $\parallel$ lines).

$\therefore AB = EC$,

and

$\angle BAC = \angle ACE$, § 128

(being homologous parts of equal $\triangle_s$).

$\therefore AB$ is $\parallel$ to EC, § 111

(two lines are $\parallel$, if the alt.-int. $\angle_s$ are equal).

$\therefore$ the figure $ABCE$ is a $\square$, § 166

(the opposite sides being parallel).

Q.E.D.

PROPOSITION XXXVI. THEOREM.

184. *The diagonals of a parallelogram bisect each other.*

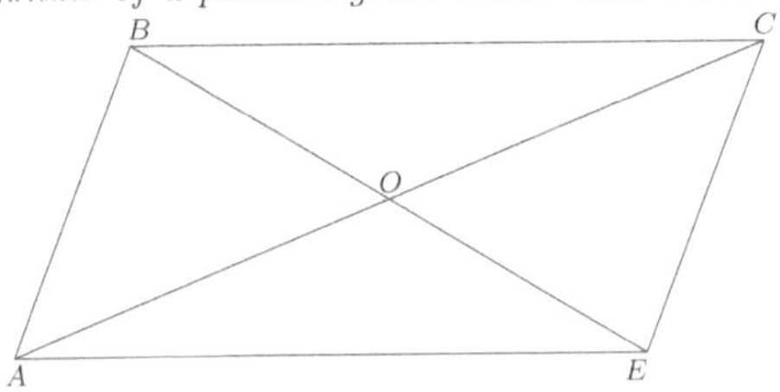

Let the figure $ABCE$ be a parallelogram, and let the diagonals AC and BE cut each other at O.

To prove that $AO = OC$, and $BO = OE$.

Proof. In the $\triangle_s AOE$ and COB,

$$AE = BC, \qquad \S\ 178$$
(*being opposite sides of a* ▱).

$$\angle OAE = \angle OCB, \qquad \S\ 110$$

$$\text{and } \angle OEA = \angle OBC,$$
(*being alt.-int.* $\angle_s$ *of* ∥ *lines*).

$$\therefore \triangle AOE = \triangle COB, \qquad \S\ 139$$
(*having two* $\angle_s$ *and the included side of the one equal, respectively, to two* $\angle_s$ *and the included side of the other*).

$$\therefore AO = OC, \text{ and } BO = OE, \qquad \S\ 128$$
(*being homologous sides of equal* $\triangle_s$).

Q.E.D.

Ex. 13. The median from the vertex to the base of an isosceles triangle is perpendicular to the base, and bisects the vertical angle.

Ex. 14. If two straight lines are cut by a transversal so that the alternate-exterior angles are equal, the two straight lines are parallel.

Ex. 15. If two parallel lines are cut by a transversal, the two exterior angles on the same side of the transversal are supplementary.

Ex. 16. If two straight lines are cut by a transversal so as to make the exterior angles on the same side of the transversal supplementary, the two lines are parallel.

PROPOSITION XXXVII. THEOREM.

185. *Two parallelograms are equal, if two sides and the included angle of the one are equal, respectively, to two sides and the included angle of the other.*

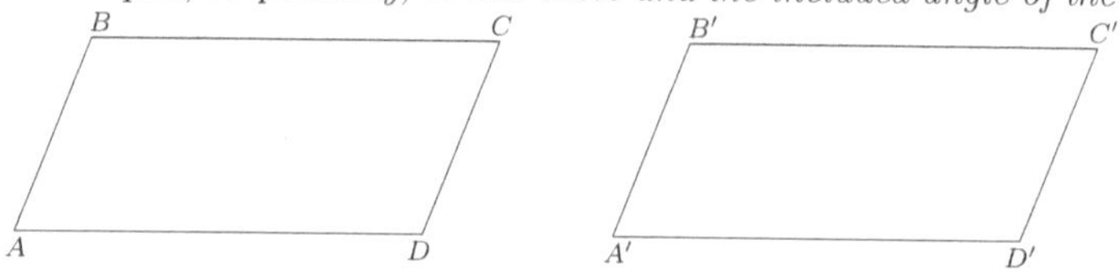

In the parallelograms $ABCD$ and $A'B'C'D'$, let AB be equal to $A'B'$, AD to $A'D'$, and angle A to A'.

To prove that the ▱s *are equal.*

Proof. Place the ▱ $ABCD$ on the ▱ $A'B'C'D'$, so that AD will fall on and coincide with its equal, $A'D'$.

Then AB will fall on $A'B'$, and B on B';
(*for* $\angle A = \angle A'$, *and* $AB = A'B'$, *by hyp.*)

Now, BC and $B'C'$ are both ∥ to $A'D'$ and drawn through B'.

∴ BC and $B'C'$ coincide, § 105
(*through a given point only one line can be drawn* ∥ *to a given line*).

Also DC and $D'C'$ are ∥ to $A'B'$ and drawn through D'.

∴ DC and $D'C'$ coincide. § 105

∴ C falls on C', § 48
(*two lines can intersect in only one point*),

∴ the two ▱s coincide, and are equal. Q.E.D.

186. COR. *Two rectangles having equal bases and altitudes are equal.*

PROPOSITION XXXVIII. THEOREM.

187. *If three or more parallels intercept equal parts on one transversal, they intercept equal parts on every transversal.*

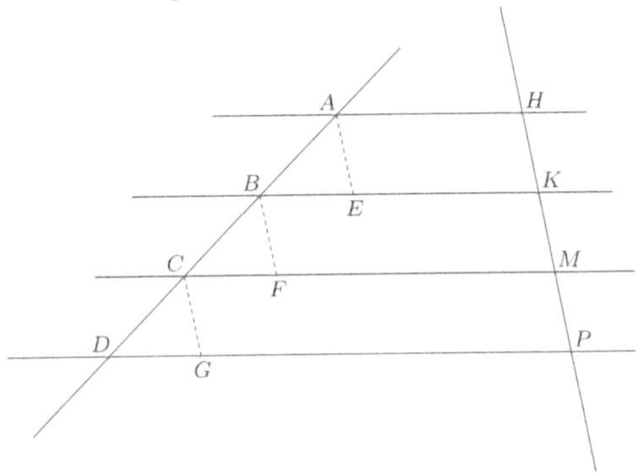

Let the parallels AH, BK, CM, DP intercept equal parts HK, KM, MP on the transversal HP.

To prove that they intercept equal parts AB, BC, CD on the transversal AD.

Proof. Suppose AH, BF, and CG drawn $\parallel$ to HP.

$\angle_s$ AEB, BFC, etc. $= \angle_s$ HKE, KMF, etc., respectively. § 112

But $\angle_s$ HKE, KMF, etc. are equal. § 112

$\therefore \angle_s$ AEB, BFC, etc. are equal. Ax. 1

Also $\angle_s$ BAE, CBF, etc. are equal. § 112

Now $AE = HK$, $BF = KM$, $CG = MP$, § 180

(parallels comprehended between parallels are equal).

$\therefore AE = BF = CG.$ Ax. 1

$\therefore \triangle ABE = \triangle BCF = \triangle CDG,$ § 139

(having two $\angle_s$ and the included side of each respectively equal).

$\therefore AB = BC = CD.$ § 128

Q.E.D.

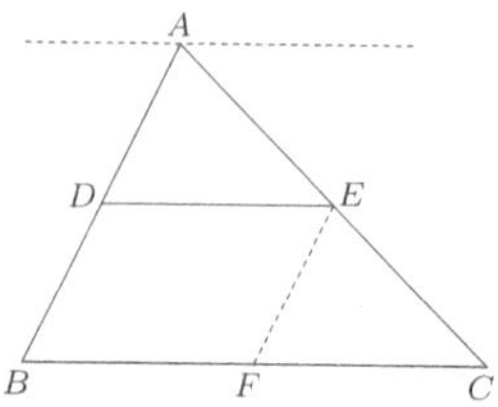

188. COR. 1. *If a line is parallel to the base of a triangle and bisects one side, it bisects the other side also.*

Let DE be $\parallel$ to BC and bisect AB. Suppose a line is drawn through A $\parallel$ to BC. Then this line is $\parallel$ to DE, by § 106. The three parallels by hypothesis intercept equal parts on the transversal AB, and therefore, by § 187, they intercept equal parts on the transversal AC; that is, the line DE bisects AC.

189. COR. 2. *The line which joins the middle points of two sides of a triangle is parallel to the third side, and is equal to half the third side.*

A line drawn through D, the middle point of AB, $\parallel$ to BC, passes through E, the middle point of AC, by § 188. Therefore the line joining D and E coincides with this parallel and is $\parallel$ to BC. Also, since EF drawn $\parallel$ to AB bisects AC, it bisects BC, by § 188; that is, $BF = FC = \frac{1}{2}BC$. But $BDEF$ is a $\square$ by § 166, and therefore $DE = BF = \frac{1}{2}BC$.

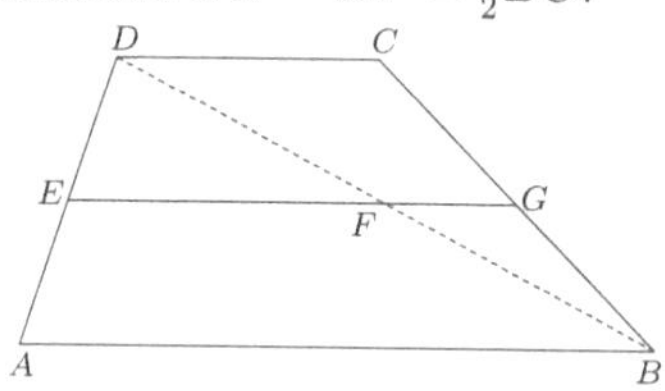

190. COR. 3. *The median of a trapezoid is parallel to the bases, and is equal to half the sum of the bases.*

Draw the diagonal DB. In the $\triangle ADB$ join E, the middle point of AD, to F, the middle point of DB. Then, by § 189, EF is $\parallel$ to AB and $= \frac{1}{2}AB$. In the $\triangle DBC$ join F to G, the middle point of BC. Then FG is $\parallel$ to DC and $= \frac{1}{2}DC$. AB and FG, being $\parallel$ to DC, are $\parallel$ to each other. But only one line can be drawn through F $\parallel$ to AB (§ 105). Therefore FG is the prolongation of EF. Hence, EFG is parallel to AB and DC, and equal to $\frac{1}{2}(AB + DC)$.

POLYGONS IN GENERAL.

191. A **polygon** is a portion of a plane bounded by straight lines.

The bounding lines are the sides, and their sum, the **perimeter** of the polygon. The angles included by the adjacent sides are the **angles** of the polygon, and the vertices of these angles are the **vertices** of the polygon. The number of sides of a polygon is evidently equal to the number of its angles.

192. A **diagonal** of a polygon is a line joining the vertices of two angles not adjacent; as, AC (Fig. 1).

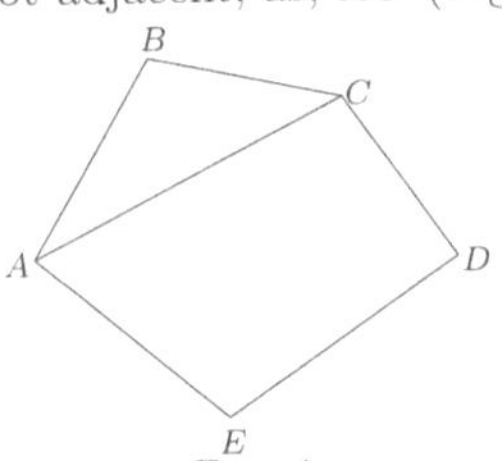

FIG. 1.

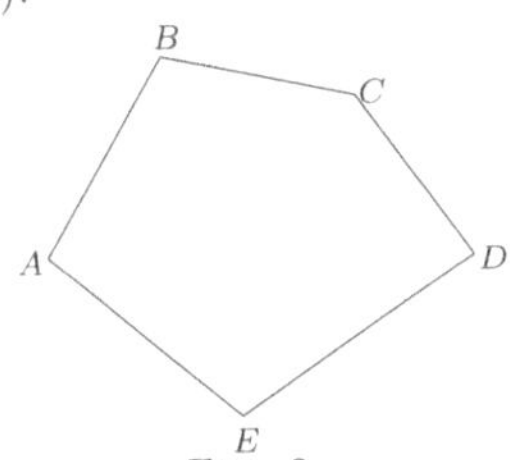

FIG. 2.

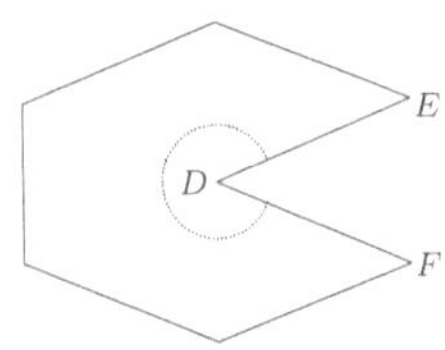

FIG. 3.

193. An **equilateral polygon** is a polygon which has all its sides equal.

194. An **equiangular polygon** is a polygon which has all its angles equal.

195. A **convex polygon** is a polygon of which no side, when produced, will enter the polygon.

196. A **concave polygon** is a polygon of which two or more sides, if produced, will enter the polygon.

197. Each angle of a convex polygon (Fig. 2) is called a *salient* angle, and is less than a straight angle.

198. The angle EDF of the concave polygon (Fig. 3) is called a *re-entrant* angle, and is greater than a straight angle.

When the term polygon is used, a *convex* polygon is meant.

199. Two polygons are *equal* when they can be divided by diagonals into the same number of triangles, equal each to each, and similarly placed; for if the polygons are applied to each other, the corresponding triangles will coincide, and hence the polygons will coincide and be equal.

200. Two polygons are *mutually equiangular*, if the angles of the one are equal to the angles of the other, each to each, when taken in the same order. Figs. 1 and 2.

201. The equal angles in mutually equiangular polygons are called *homologous* angles; and the sides which are included by homologous angles are called *homologous* sides.

202. Two polygons are *mutually equilateral*, if the sides of the one are equal to the sides of the other, each to each, when taken in the same order. Figs. 1 and 2.

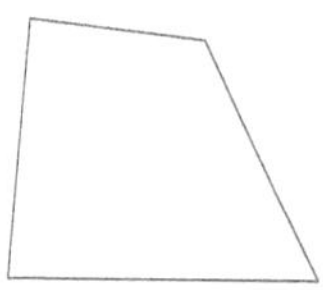

FIG. 4.

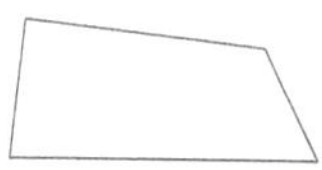

FIG. 5.

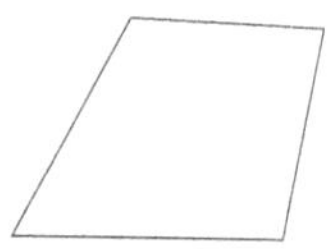

FIG. 6.

FIG. 7.

203. Two polygons may be mutually equiangular without being mutually equilateral; as, Figs. 4 and 5.

And, *except in the case of triangles*, two polygons may be mutually equilateral without being mutually equiangular; as, Figs. 6 and 7.

If two polygons are mutually equilateral and mutually equiangular *they are equal*, for they can be made to coincide.

204. A polygon of three sides is called a *triangle*; one of four sides, a *quadrilateral*; one of five sides, a *pentagon*; one of six sides, a *hexagon*; one of seven sides, a *heptagon*; one of eight sides, an *octagon*; one of ten sides, a *decagon*; one of twelve sides, a *dodecagon*.

PROPOSITION XXXIX. THEOREM.

205. *The sum of the interior angles of a polygon is equal to two right angles, taken as many times less two as the figure has sides.*

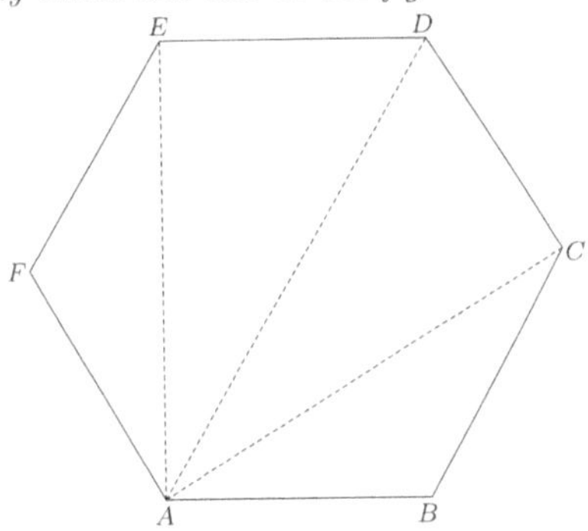

Let the figure $ABCDEF$ be a polygon, having n sides.

To prove that $\angle A + \angle B + \angle C$, *etc.* $= (n-2)2$ *rt.* $\angle_s$.

Proof. From A draw the diagonals AC, AD, and AE.

The sum of the $\angle_s$ of the $\triangle_s$ is equal to the sum of the $\angle_s$ of the polygon.

Now, there are $(n-2)$ $\triangle_s$,

and the sum of the $\angle_s$ of each $\triangle = 2$ rt. $\angle_s$. § 129

$\therefore$ the sum of the $\angle_s$ of the $\triangle_s$, that is, the sum of the $\angle_s$ of the polygon is equal to $(n-2)2$ rt. $\triangle_s$. Q.E.D.

206. COR. *The sum of the angles of a quadrilateral equals 4 right angles; and if the angles are all equal, each is a right angle. In general, each angle of an equiangular polygon of n sides is equal to* $\dfrac{2(n-2)}{n}$ *right angles.*

Ex. 17. How many diagonals can be drawn in a polygon of n sides?

PROPOSITION XL. THEOREM.

207. *The exterior angles of a polygon, made by producing each of its sides in succession, are together equal to four right angles.*

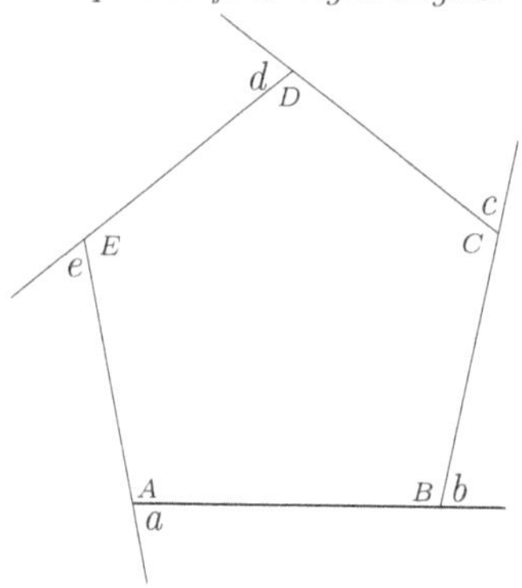

Let the figure *ABCDE* be a polygon, having its sides produced in succession.

To prove the sum of the ext. $\angle_s = 4$ *rt.* $\angle_s$.

Proof. Denote the int. $\angle_s$ of the polygon by A, B, C, D, E, and the corresponding ext. $\angle_s$ by a, b, c, d, e.

$$\angle A + \angle a = 2 \text{ rt. } \angle_s, \qquad \text{§ 89}$$

and

$$\angle B + \angle b = 2 \text{ rt. } \angle_s,$$
$$(\textit{being sup.-adj. } \angle_s).$$

In like manner each pair of adj. $\angle_s = 2$ rt. $\angle_s$.

$\therefore$ the sum of the interior and exterior $\angle_s$ of a polygon of n sides is equal to $2n$ rt. $\angle_s$.

But the sum of the interior $\angle_s = (n-2)2$ rt. $\angle_s$ § 205
$= 2n$ rt. $\angle_s - 4$ rt. $\angle_s$.

$\therefore$ the sum of the exterior $\angle_s = 4$ rt. $\angle_s$. Q.E.D.

Ex. 18. How many sides has a polygon if the sum of its interior $\angle_s$ is twice the sum of its exterior $\angle_s$? ten times the sum of its exterior $\angle_s$?

SYMMETRY.

208. Two points are said to be **symmetrical** with respect to a third point, called the **centre of symmetry**, if this third point bisects the straight line which joins them.

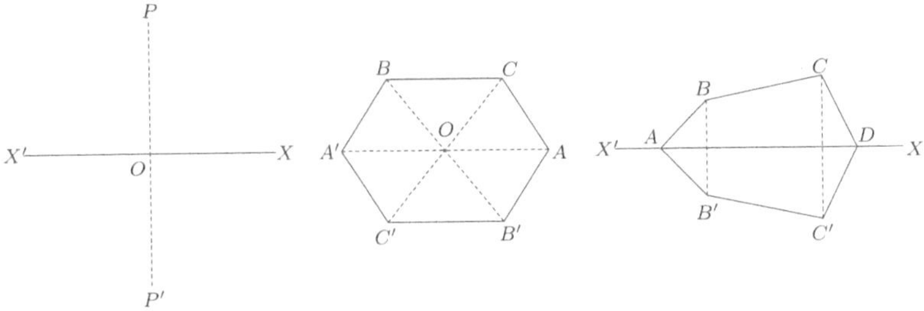

Two points are said to be *symmetrical* with respect to a straight line, called the **axis of symmetry**, if this straight line bisects at right angles the straight line which joins them.

Thus, P and P' are symmetrical with respect to O as a centre, and XX' as an axis, if O bisects the line PP', and if XX' bisects PP' at right angles.

209. A figure is symmetrical with respect to a point as a centre of symmetry, if the point bisects every straight line drawn through it and terminated by the boundary of the figure.

210. A figure is symmetrical with respect to a line as an axis of symmetry if one of the parts of the figure coincides, point for point, with the other part when it is folded over on that line as an axis.

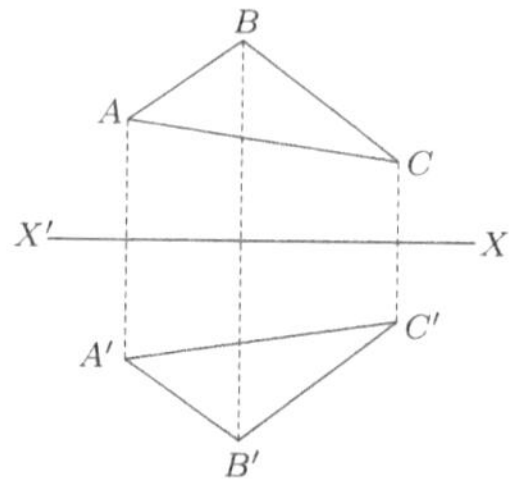

211. Two figures are said to be symmetrical with respect to an axis if every point of one has a corresponding symmetrical point in the other.

Thus, if every point in the figure $A'B'C'$ has a symmetrical point in ABC, with respect to XX' as an axis, the figure $A'B'C'$ is symmetrical to ABC with respect to XX' as an axis.

PROPOSITION XLI. THEOREM.

212. *A quadrilateral which has two adjacent sides equal, and the other two sides equal, is symmetrical with respect to the diagonal joining the vertices of the angles formed by the equal sides, and the diagonals are perpendicular to each other.*

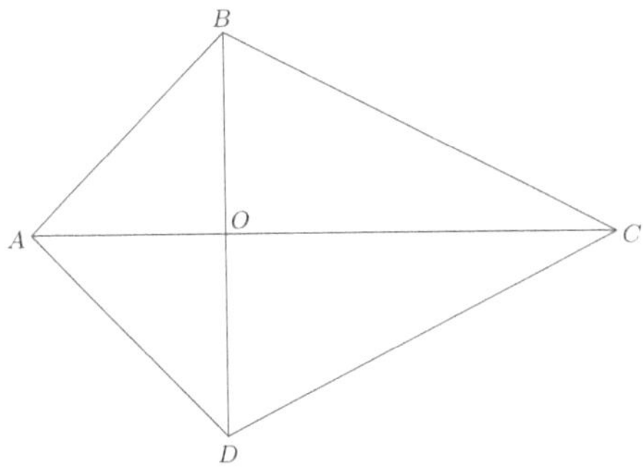

Let $ABCD$ be a quadrilateral, having AB equal to AD, and CB equal to CD, and having the diagonals AC and BD.

To prove that the diagonal AC is an axis of symmetry, and that it is $\perp$ to the diagonal BD.

Proof. In the $\triangle_s ABC$ and ADC,

$$AB = AD, \text{ and } BC = DC, \qquad \text{Hyp.}$$

and

$$AC = AC. \qquad \text{Iden.}$$

$$\therefore \triangle ABC = \triangle ADC. \qquad \S\ 150$$

$$\therefore \angle BAC = \angle DAC, \text{ and } \angle BCA = \angle DCA.$$

Hence, if ABC is turned on AC as an axis until it falls on ADC, AB will fall upon AD, CB on CD, and OB on OD.

$$\therefore \text{ the } \triangle ABC \text{ will coincide with the } \triangle ADC.$$

$$\therefore AC \text{ is an axis of symmetry } (\S\ 210) \text{ and is } \perp \text{ to } BD. \qquad \S\ 208$$

Q.E.D.

Proposition XLII. Theorem.

213. *If a figure is symmetrical with respect to two axes perpendicular to each other, it is symmetrical with respect to their intersection as a centre.*

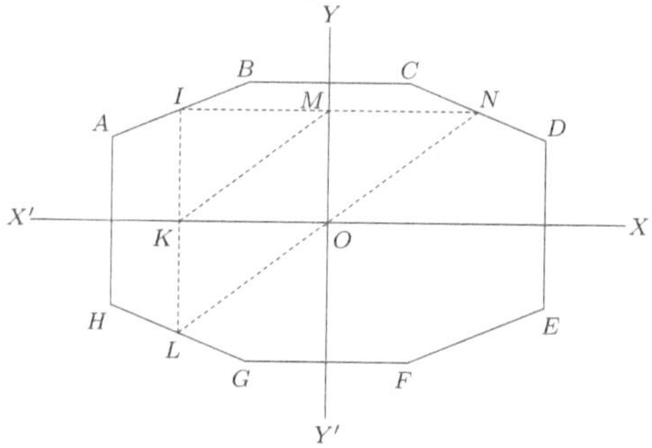

Let the figure $ABCDEFGH$ be symmetrical with respect to the two perpendicular axes XX', YY', which intersect at O.

To prove that O is the centre of symmetry of the figure.

Proof. Let N be any point in the perimeter.

Suppose NMI drawn $\perp$ to YY', $IKL \perp$ to XX'.

Then

NI is $\parallel$ to XX' and IL is $\parallel$ to YY'. § 104

Draw LO, ON, and KM.

Now

$KI = KL$, § 208

(*the figure being symmetrical with respect to XX'*).

But

$KI = OM$. § 180

$\therefore KL = OM$, and $KLOM$ is a $\square$. § 183

$\therefore LO$ is equal and parallel to KM. § 183

In like manner ON is equal and parallel to KM.

$\therefore LON$ is a straight line. § 105

$\therefore O$ bisects LN, *any* straight line and therefore *every* straight line drawn through O and terminated by the perimeter.

$\therefore O$ is the centre of symmetry of the figure. Q.E.D.

REVIEW QUESTIONS ON BOOK I.

1. What is the subject-matter of Geometry?
2. What is a geometric magnitude?
3. What is an axiom? a theorem? a converse theorem? an opposite theorem? a contradictory theorem?
4. Define a straight line; a curved line; a broken line; a plane surface; a curved surface.
5. How many points are necessary to determine a straight line?
6. How many straight lines are necessary to determine a point?
7. On what does the magnitude of an angle depend?
8. Define a straight angle; a right angle; an oblique angle.
9. Define adjacent angles; complementary angles; supplementary angles; conjugate angles.
10. Define parallel lines and give the axiom of parallels.
11. If two lines in the same plane are parallel and cut by a transversal, what pairs of angles are equal? what pairs are supplementary?
12. Define a right triangle; an isosceles triangle; a scalene triangle.
13. To how many right angles is the sum of the angles of a triangle equal? the sum of the acute angles of a right triangle?
14. To what angles is the exterior angle of a triangle equal?
15. What is the test of equality of two geometric magnitudes?
16. How does a reciprocal theorem differ from a converse theorem?
17. State the three cases in which two triangles are equal.
18. State the cases in which two right triangles are equal.
19. What is meant by a locus of points?
20. Where are the points located in a plane that are each equidistant from two given points? from two intersecting lines?
21. Define a parallelogram; a trapezoid; an isosceles trapezoid.
22. When is a figure symmetrical with respect to a centre?

23. When is a figure symmetrical with respect to an axis?

24. Must a triangle be equiangular if equilateral? must a triangle be equilateral if equiangular?

25. When are two polygons said to be mutually equiangular?

26. When are two polygons said to be mutually equilateral?

27. Can two polygons of more than three sides be mutually equiangular without being mutually equilateral? mutually equilateral without being mutually equiangular?

28. What line do two points each equidistant from the extremities of a given straight line determine?

METHODS OF PROVING THEOREMS.

214. There are *three* general methods of proving theorems, the **synthetic**, the **analytic**, and the **indirect** methods.

The *synthetic* method is the method employed in most of the theorems already given, and consists in putting together known truths in order to obtain a new truth.

The *analytic* method is the reverse of the synthetic method. It asserts that the conclusion is true if another proposition is true, and so on step by step, until a known truth is reached. Thus, proposition A is true if proposition B is true, and B is true if C is true; but C *is* true, hence A and B are true.

If a known truth *suggests* the required proof, it is best to use the synthetic form at once. If no proof occurs to the mind, it is necessary to use the analytic method to *discover* the proof, and then the synthetic proof may be given.

The *indirect* method, or the method of *reductio ad absurdum*, is illustrated on page 45. It consists in proving a theorem to be true by proving its contradictory to be false.

215. Generally auxiliary lines are required, as a line *connecting two points*; a line *parallel to or perpendicular to a given line*; a line *produced by its own length*; a line *making with another line an angle equal to a given angle.*

Two lines are proved equal by proving them *homologous sides of equal triangles*; or *legs of an isosceles triangle*; or *opposite sides of a parallelogram.*

Two angles are proved equal by proving them *alternate-interior angles or exterior-interior angles of parallel lines*; or *homologous angles of equal triangles*; or *base angles of an isosceles triangle*; or *opposite angles of a parallelogram.*

Two suggestions are of special importance to the beginner:

1. *Draw as accurate figures as possible.*
2. *Draw as general figures as possible.*

EXERCISES.

Prove by the analytic method:

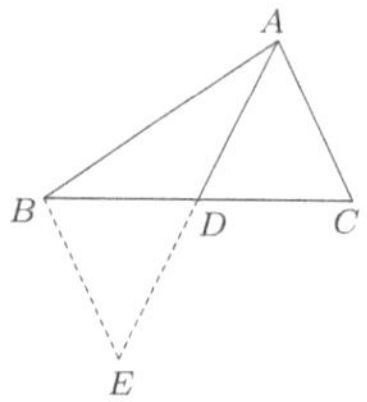

Ex. 19. *A median of a triangle is less than half the sum of the two adjacent sides.*

To prove the median $AD < \frac{1}{2}(AB + AC)$.

Now

$$AD < \tfrac{1}{2}(AB + AC),$$

if

$$2AD < AB + AC.$$

This suggests producing AD by its own length to E, and joining BE.

Then

$$AE = 2AD,$$

and

$$2AD < AB + AC \text{ if } AE < AB + AC.$$

But

$$AE < AB + BE. \qquad \text{§ 138}$$

$$\therefore AE < AB + AC \text{ if } AC = BE.$$

And

$$AC = BE \text{ if } \triangle ACD = \triangle EBD. \qquad \text{§ 128}$$

But

$$\triangle ACD = \triangle EBD. \qquad \text{§ 143}$$

For

$$CD = DB, \qquad \text{Hyp.}$$

$$AD = DE, \qquad \text{Const.}$$

and

$$\angle ADC = \angle BDE. \qquad \text{§ 93}$$

$$\therefore AE < AB + AC.$$

$$\therefore AD < \tfrac{1}{2}(AB + AC).$$

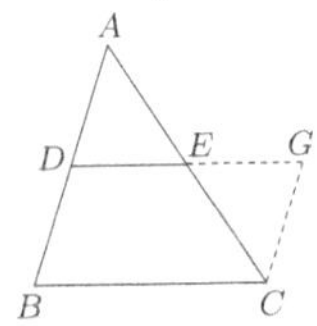

Ex. 20. *A straight line which bisects two sides of a triangle is parallel to the third side.*

If $AD = DB$ and $AE = EC$, to prove $DE \parallel$ to BC.

Draw $CG \parallel$ to BA, and produce DE to meet it at G.

DE is $\parallel$ to BC if $BCGD$ is a $\square$. § 166

$BCGD$ is a $\square$ if $CG = BD$. § 183

$CG = BD$ if each is equal to AD. Ax. 1

Now $BD = AD$. Hyp.

And $CG = AD$ if $\triangle CGE = \triangle ADE$. § 128

But $\triangle CGE = \triangle ADE$. § 139

For $EC = AE$. Hyp.

$\angle GEC = \angle AED$. § 93

$\angle ECG = \angle DAE$. § 110

$\therefore DE$ is $\parallel$ to BC.

Prove by the synthetic method:

Ex. 21. *The middle point of the hypotenuse of a right triangle is equidistant from the three vertices.*

From D, the middle point, draw $DE \perp$ to CB.

DE is $\parallel$ to AC (why?), and DE bisects CB (why?).

$\therefore D$ is equidistant from B, A, and C. (Why?)

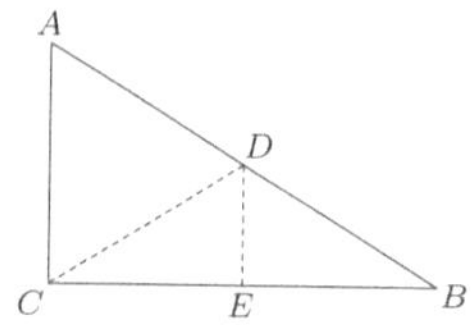

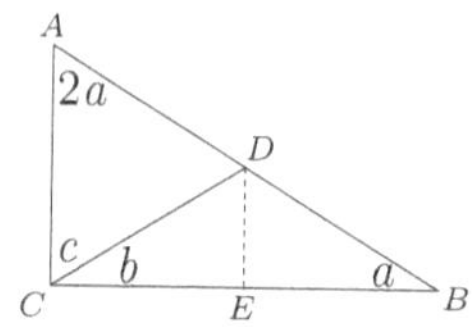

Ex. 22. *If one acute angle of a right triangle is double the other, the hypotenuse is double the shorter leg.*

The median $CD = BD = AD$ (Ex. 21).

Then $\angle b = \angle a$; and $\angle c = \angle 2a$. (Why?)

Now $a + 2a = 90°$. (Why?)

$\therefore \angle a = 30°$; $\angle 2a = 60°$; $\angle c = 60°$.

$\therefore \triangle ACD$ is equilateral (why?), and AD, half of $AB = AC$. $\therefore AB = 2AC$.

Ex. 23. *If two triangles have two sides of the one equal, respectively, to two sides of the other, and the angles opposite two equal sides equal, the angles opposite the other two equal sides are equal or supplementary, and if equal the triangles are equal.*

Let $AC = A'C'$, $BC = B'C'$, and $\angle B = \angle B'$.

Place $\triangle A'B'C'$ on $\triangle ABC$ so that $B'C'$ shall coincide with BC, and $\angle A'$ and $\angle A$ shall be on the same side of BC.

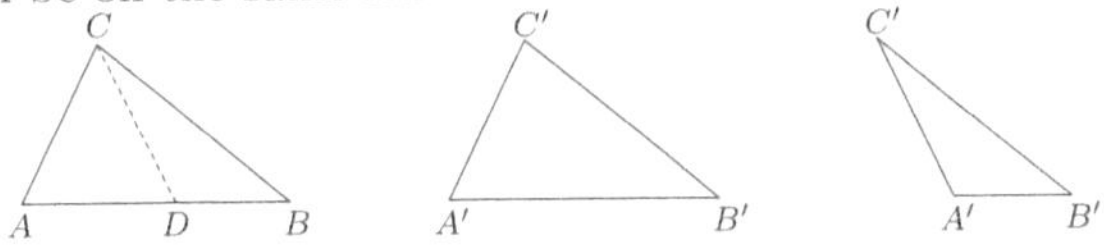

Since $\angle B' = \angle B$, $B'A'$ will fall along BA, and A' will fall at A or at some other point in BA, as D. If A' falls at A, the $\triangle_s A'B'C'$ and ABC coincide and are equal.

If A' falls at D, the $\triangle_s A'B'C'$ and DBC coincide and are equal.

Since $CD = C'A' = CA$, $\angle A = \angle CDA$. (Why?)

But $\angle_s CDA$ and CDB are supplements. (Why?)

$\therefore \angle_s A$ and CDB are supplements. (Why?)

Draw figures and show that the triangles are equal:

1. If the given angles B and B' are both right or both obtuse angles.

2. If the required angles A and A' are both acute, both right, or both obtuse.

3. If AC and $A'C'$ are not less than BC and $B'C'$, respectively.

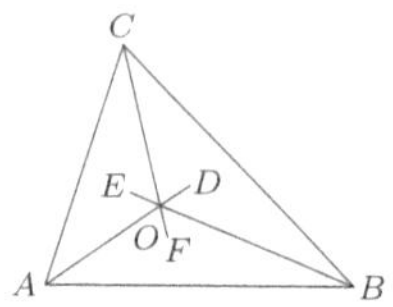

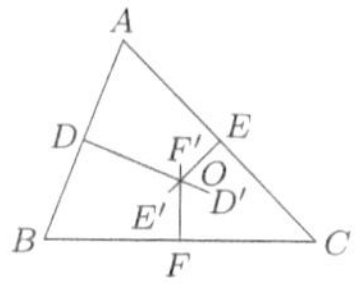

Ex. 24. *The bisectors of the angles of a triangle meet in a point which is equidistant from the sides of the triangle.*

Let the bisectors AD and BE intersect at O. Then O being in AD is equidistant from AC and AB. (Why?) And O being in BE is equidistant from BC and AB. Hence, O is equidistant from AC and BC, and therefore in the bisector CF. (Why?)

Ex. 25. *The perpendicular bisectors of the sides of a triangle meet in a point which is equidistant from the vertices of the triangle.*

Let the $\perp$ bisectors EE' and DD' intersect at O. Then O being in EE' is equidistant from A and C. (Why?) And O being in DD' is equidistant from A and B. Hence, O is equidistant from B and C, and therefore is in the $\perp$ bisector FF'. (Why?)

Ex. 26. *The perpendiculars from the vertices of a triangle to the opposite sides meet in a point.*

Let the $\perp_s$ be AH, BP, and CK. Through A, B, C suppose $B'C'$, $A'C'$, $A'B'$, drawn $\parallel$ to BC, AC, AB, respectively. Then AH is $\perp$ to $B'C'$. (Why?) Now $ABCB'$ and $ACBC'$ are $\square_s$ (why?) and $AB' = BC$, and $AC' = BC$. (Why?) That is, A is the middle point of $B'C'$. In the same way, B and C are the middle points of $A'C'$ and $A'B'$, respectively. Therefore, AH, BP, and CK are the $\perp$ bisectors of the sides of the $\triangle A'B'C'$. Hence, they meet in a point. (Why?)

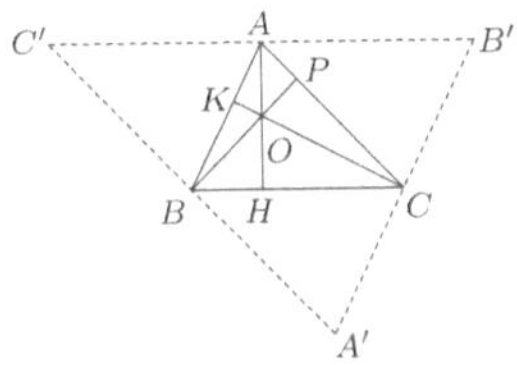

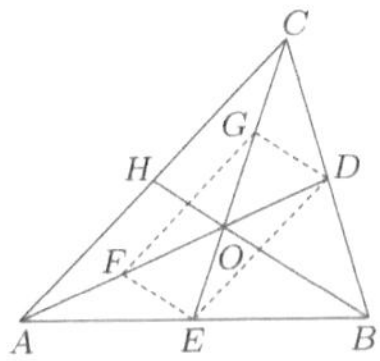

Ex. 27. *The medians of a triangle meet in a point which is two thirds of the distance from each vertex to the middle of the opposite side.*

Let the two medians AD and CE meet in O. Take F the middle point of OA, and G of OC. Join GF, FE, ED, and DG. In $\triangle AOC$, GF is $\|$ to AC and equal to $\frac{1}{2}AC$. (Why?) DE is $\|$ to AC and equal to $\frac{1}{2}AC$. (Why?) Hence, $DGFE$ is a $\square$. (Why?) Hence, $AF = FO = OD$, and $CG = GO = OE$. (Why?) Hence, *any median* cuts off *on any other median* two thirds of the distance from the vertex to the middle of the opposite side. Therefore, the median from B will cut off AO, two thirds of AD; that is, will pass through O.

NOTE. If *three* or more lines pass through the same point, they are called *concurrent* lines.

Ex. 28. If an angle is bisected, and if a line is drawn through the vertex perpendicular to the bisector, this line forms equal angles with the sides of the given angle.

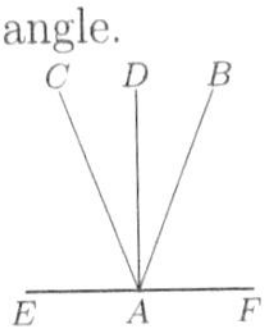

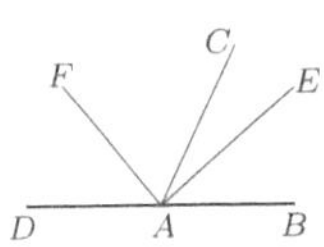

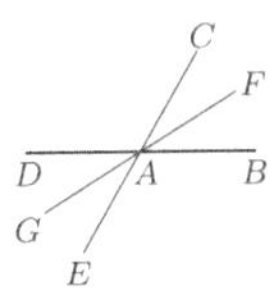

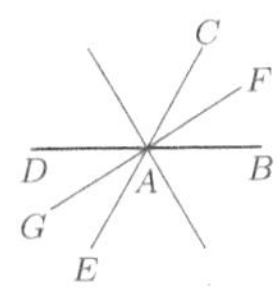

Ex. 29. The bisectors of two supplementary adjacent angles are perpendicular to each other.

Ex. 30. If the bisectors of two adjacent angles are perpendicular to each other, the adjacent angles are supplementary.

Ex. 31. The bisector of one of two vertical angles bisects the other.

Ex. 32. The bisectors of two vertical angles form one line.

Ex. 33. The bisectors of the two pairs of vertical angles formed by two intersecting lines are perpendicular to each other.

Ex. 34. The bisector of the vertical angle of an isosceles triangle bisects the base, and is perpendicular to the base.

$$\triangle ADC = \triangle BDC \text{ (§ 143)}$$

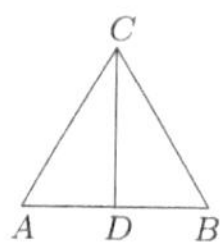

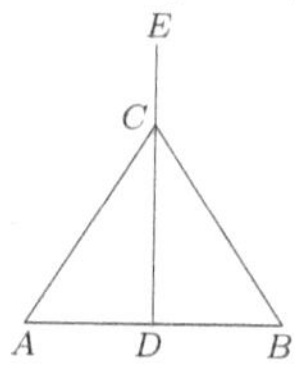

Ex. 35. The perpendicular bisector of the base of an isosceles triangle passes through the vertex and bisects the angle at the vertex (§ 160).

Ex. 36. If the perpendicular bisector of the base of a triangle passes through the vertex, the triangle is isosceles.

Ex. 37. Any point in the bisector of the vertical angle of an isosceles triangle is equidistant from the extremities of the base (Ex. 34, § 160).

Ex. 38. If the bisector of an angle of a triangle is perpendicular to the opposite side, the triangle is isosceles.

Ex. 39. If two isosceles triangles are on the same base, a straight line passing through their vertices is perpendicular to the base, and bisects the base (§ 161).

Ex. 40. Two isosceles triangles are equal when a side and an angle of the one are equal, respectively, to the homologous side and angle of the other.

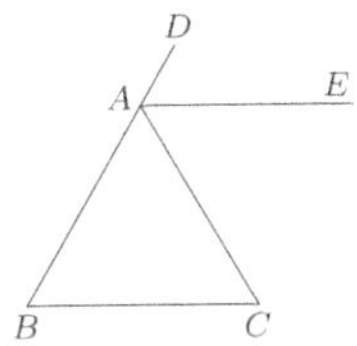

Ex. 41. The bisector of an exterior angle of an isosceles triangle, formed by producing one of the legs through the vertex, is parallel to the base. Why does $\angle DAC = \angle B + \angle C$? Why is $\angle DAE = \angle ABC$? Why is $AE \parallel$ to BC?

Ex. 42. If the bisector of an exterior angle of a triangle is parallel to one side, the triangle is isosceles.

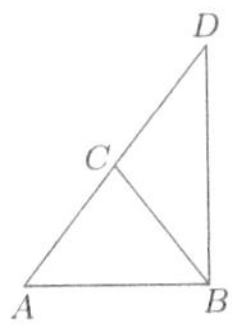

Ex. 43. If one of the legs of an isosceles triangle is produced through the vertex by its own length, the line joining the end of the leg produced to the nearer end of the base is perpendicular to the base.

$$\angle CBA = \angle A, \text{ and } \angle CBD = \angle D. \text{ (Why?)}$$

$$\therefore \angle ABD = \angle A + \angle D.$$

Ex. 44. A line drawn from the vertex of the right angle of a right triangle to the middle point of the hypotenuse divides the triangle into two isosceles triangles.

Ex. 45. If the equal sides of an isosceles triangle are produced through the vertex so that the external segments are equal, the extremities of these segments will be equally distant from the extremities of the base, respectively.

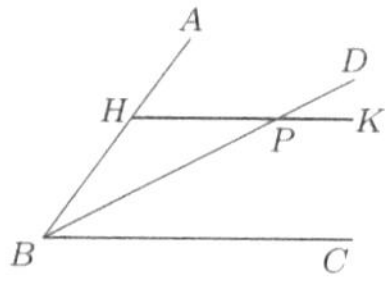

Ex. 46. If through any point in the bisector of an angle a line is drawn parallel to either of the sides of the angle, the triangle thus formed is isosceles.

Ex. 47. Through any point C in the line AB an intersecting line is drawn, and from any two points in this line equidistant from C perpendiculars are dropped on AB or AB produced. Prove that these perpendiculars are equal.

Ex. 48. If the median drawn from the vertex of a triangle to the base is equal to half the base, the vertical angle is a right angle.

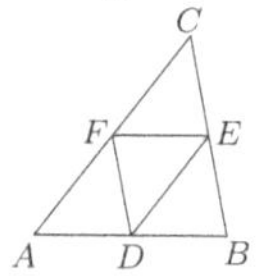

Ex. 49. The lines joining the middle points of the sides of a triangle divide the triangle into four equal triangles.

Ex. 50. The altitudes upon the legs of an isosceles triangle are equal.

Rt. $\triangle BEC =$ rt. $\triangle CDB$ (§ 141).

Ex. 51. If the altitudes upon two sides of a triangle are equal, the triangle is isosceles.

Rt. $\triangle BEC =$ rt. $\triangle CDB$ (§ 151).

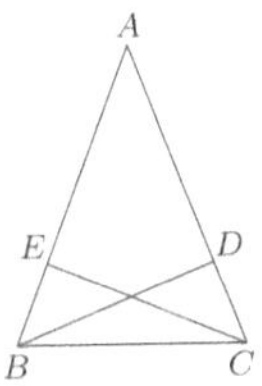

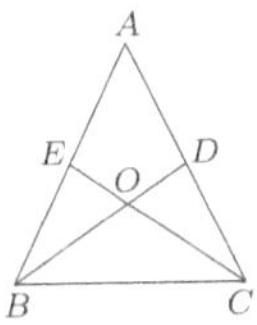

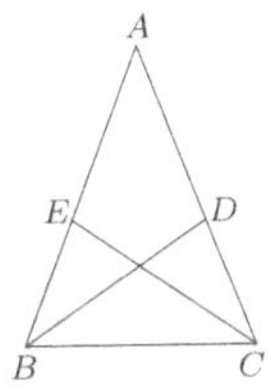

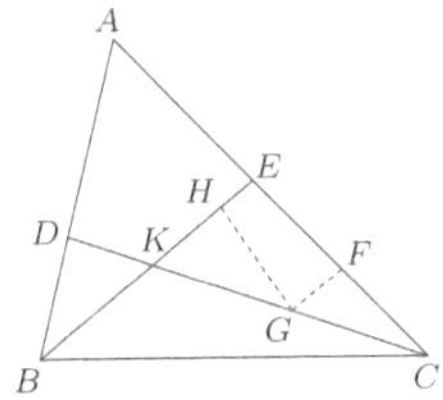

Ex. 52. The medians drawn to the legs of an isosceles triangle are equal.

$$\triangle BEC = \triangle CDB \text{ (§ 143).}$$

Ex. 53. If the medians to two sides of a triangle are equal, the triangle is isosceles.

$$BO = CO, \text{ and } OE = OD \text{ (Ex. 27).}$$

$$\angle BOE = \angle COD. \quad \therefore \triangle BOE = \triangle COD \text{ (§ 143).}$$

Ex. 54. The bisectors of the base angles of an isosceles triangle are equal.

$$\triangle BEC = \triangle CDB \text{ (§ 139).}$$

Ex. 55. Opposite Theorem. If a triangle is not isosceles, the bisectors of the base angles are not equal.

Let $\angle ABC$ be greater than $\angle ACB$; then $KC > KB$. (Why?)

Now $CD > BE$, if KD is greater than or equal to KE.

But suppose $KD < KE$. Lay off $KH = KD$ and $KG = KB$, join HG, and draw $GF \parallel$ to BE.

$\triangle KDB = \triangle KHG$. (Why?) $\therefore \angle KHG = \angle KDB$. (Why?)

$\therefore \angle KEC$ is greater than $\angle KHG$. (Why?) $\therefore GF > HE$. (Why?)

$\angle GFC$ is greater than $\angle FCG$ ($\frac{1}{2}ACB$). $\therefore CG > GF$, and $> HE$.

$\therefore KC - KG > KE - KH$, or $KC + KD > KB + KE$, or $CD > BE$.

Ex. 56. State the converse theorem of Ex. 54. Is the converse theorem true?

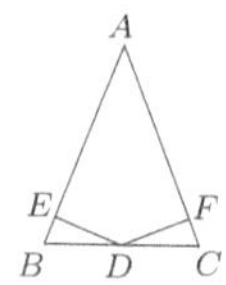

Ex. 57. The perpendiculars dropped from the middle point of the base upon the legs of an isosceles triangle are equal.

$$\triangle BED = \triangle CFD \text{ (§ 141)}.$$

Ex. 58. State and prove the converse.

$$\triangle BED = \triangle CFD \text{ (§ 151)}.$$

Ex. 59. The difference of the distances from any point in the base produced of an isosceles triangle to the equal sides of the triangle is constant.

Rt. $\triangle DGC =$ rt. $\triangle DFC$. (Why?) $\therefore DF = DG$.

$\therefore DE - DF = DE - DG = EG$, the $\perp$ distance between the two $\|_s$, BA and CH.

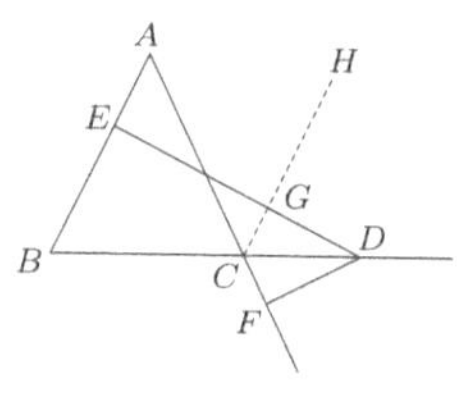

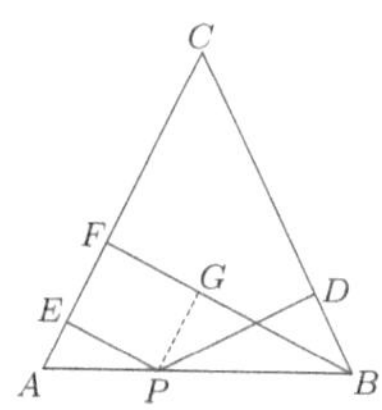

Ex. 60. The sum of the perpendiculars dropped from any point in the base of an isosceles triangle to the legs is constant, and equal to the altitude upon one of the legs.

Let PE and PD be the $\perp_s$ and BF the altitude.

Draw $PG \perp$ to BF.

$EPGF$ is a parallelogram. (Why?) $\therefore GF = PE$. It remains to prove $GB = PD$.

The rt. $\triangle PGB =$ the rt. $\triangle BDP$. (Why?)

Ex. 61. The sum of the perpendiculars dropped from any point within an equilateral triangle to the three sides is constant, and equal to the altitude.

AD is the altitude, PE, PG, and PF the three perpendiculars. Through P draw $HK \parallel$ to BC, meeting AD at M.

Then

$$MD = PE. \text{ (Why?)}$$

$$PG + PF = AM \text{ (Ex. 60).}$$

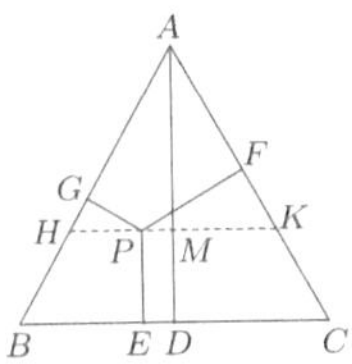

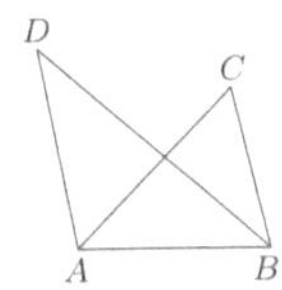

Ex. 62. ABC and ABD are two triangles on the same base AB, and on the same side of it, the vertex of each triangle being without the other. If AC equals AD, show that BC cannot equal BD (§ 154).

Ex. 63. The sum of the lines which join a point within a triangle to the three vertices is less than the perimeter, but greater than half the perimeter.

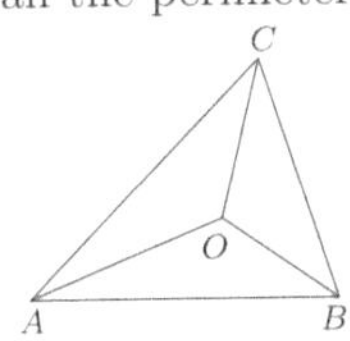

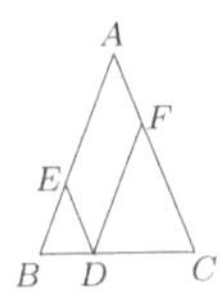

Ex. 64. If from any point in the base of an isosceles triangle parallels to the legs are drawn, a parallelogram is formed whose perimeter is constant, being equal to the sum of the legs of the triangle.

Ex. 65. The bisector of the vertical angle A of a triangle ABC, and the bisectors of the exterior angles at the base formed by producing the sides AB and AC, meet in a point which is equidistant from the base and the sides produced (§ 162).

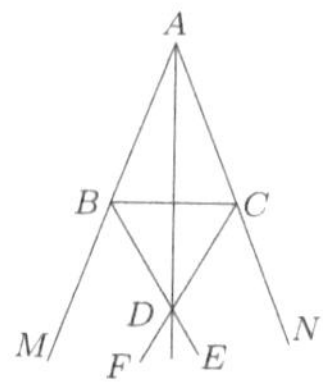

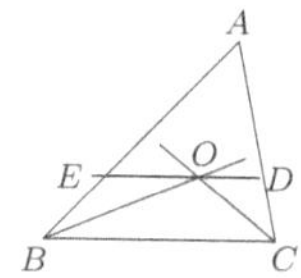

Ex. 66. If the bisectors of the base angles of a triangle are drawn, and through their point of intersection a line is drawn parallel to the base, the length of this parallel between the sides is equal to the sum of the segments of the sides between the parallel and the base.

$$\angle EOB = \angle OBC = \angle OBE. \quad \therefore BE = EO.$$

Ex. 67. The bisector of the vertical angle of a triangle makes with the perpendicular from the vertex to the base an angle equal to half the difference of the base angles.

Let $\angle B$ be greater than $\angle A$.

$$\angle DCE = 90° - \angle A - \angle ACD.$$

$$\angle ACD = 90° - \tfrac{1}{2}\angle A - \tfrac{1}{2}\angle B.$$

$$\therefore \angle DCE = 90° - \angle A - (90° - \tfrac{1}{2}\angle A - \tfrac{1}{2}\angle B) = \tfrac{1}{2}\angle B - \tfrac{1}{2}\angle A.$$

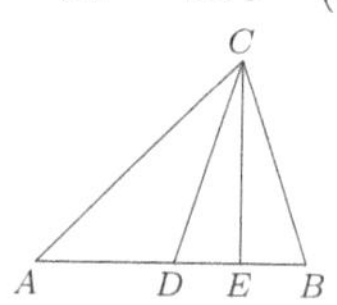

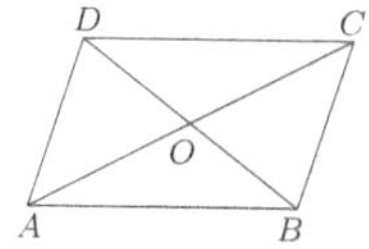

Ex. 68. If the diagonals of a quadrilateral bisect each other, the figure is a parallelogram.

Prove $\triangle AOB = \triangle COD$.

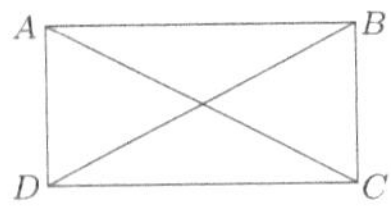

Ex. 69. The diagonals of a rectangle are equal.

Prove $\triangle ABC = \triangle BAD$.

Ex. 70. If the diagonals of a parallelogram are equal, the figure is a rectangle.

Ex. 71. The diagonals of a rhombus are perpendicular to each other, and bisect the angles of the rhombus.

Ex. 72. The diagonals of a square are perpendicular to each other, and bisect the angles of the square.

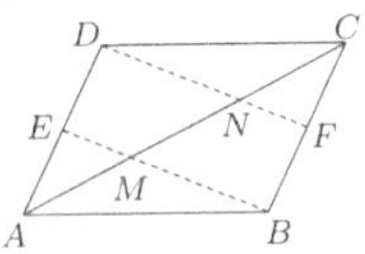

Ex. 73. Lines from two opposite vertices of a parallelogram to the middle points of the opposite sides trisect the diagonal.

$EBFD$ is a ▱ (why?), and DF is $\parallel$ to EB.

$AM = MN$, and $MN = CN$ (§ 188).

Ex. 74. The lines joining the middle points of the sides of any quadrilateral, taken in order, enclose a parallelogram.

Prove HG and $EF \parallel$ to AC; and FG and $EH \parallel$ to BD (§ 189).

Then HG and EF are each equal to $\frac{1}{2}AC$.

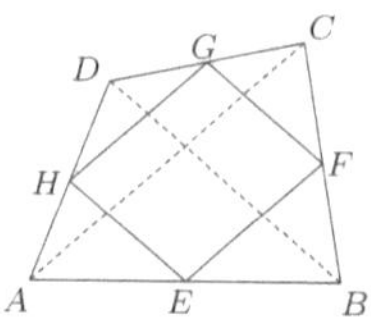

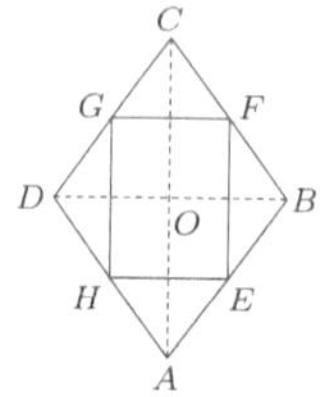

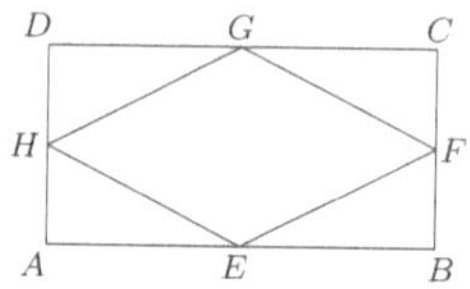

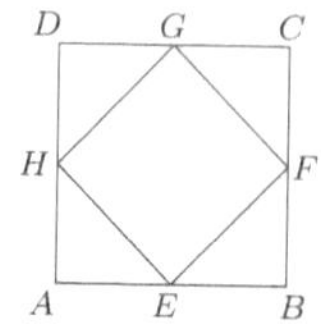

Ex. 75. The lines joining the middle points of the sides of a rhombus, taken in order, enclose a rectangle. (Proof similar to that of Ex. 74.)

Ex. 76. The lines joining the middle points of the sides of a rectangle (not a square), taken in order, enclose a rhombus.

Ex. 77. The lines joining the middle points of the sides of a square, taken in order, enclose a square.

Ex. 78. The lines joining the middle points of the sides of an isosceles trapezoid, taken in order, enclose a rhombus or a square.

SHR and QFP drawn $\perp$ to AB are parallel. $\therefore PQSR$ is a $\square$, and by Const. is a rectangle or a square.

$\therefore EFGH$ is a rhombus or a square (Exs. 76, 77).

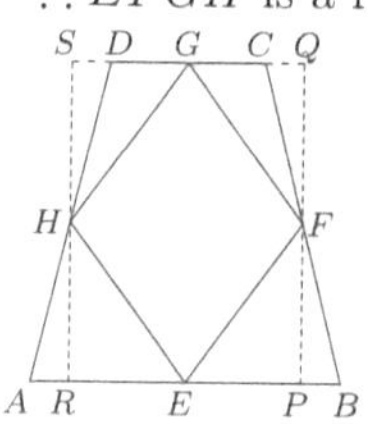

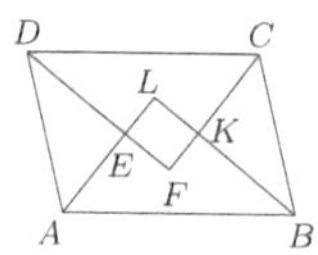

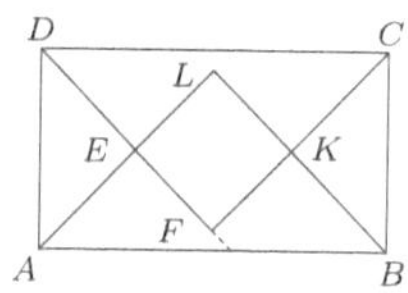

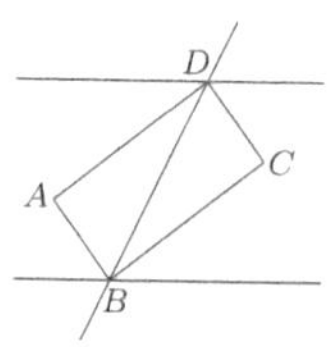

Ex. 79. The bisectors of the angles of a rhomboid enclose a rectangle.

Ex. 80. The bisectors of the angles of a rectangle enclose a square.

Ex. 81. If two parallel lines are cut by a transversal, the bisectors of the interior angles form a rectangle.

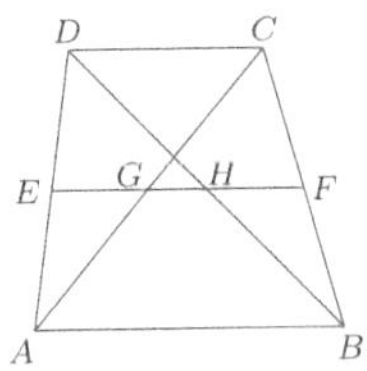

Ex. 82. The median of a trapezoid passes through the middle points of the two diagonals.

The median EF is $\parallel$ to AB and bisects AD (§ 190).

$\therefore$ it bisects DB.

Likewise EF bisects BC and BD.

Ex. 83. The lines joining the middle points of the diagonals of a trapezoid is equal to half the difference of the bases.

$$\triangle BFG = \triangle DFC. \text{ (Why?) } \therefore EF = \tfrac{1}{2}AG \text{ (§ 180)}.$$

$$CF = FG,\ DC = BG.$$

$$\therefore AG = AB - DC.\ \therefore EF = \tfrac{1}{2}(AB - DC)$$

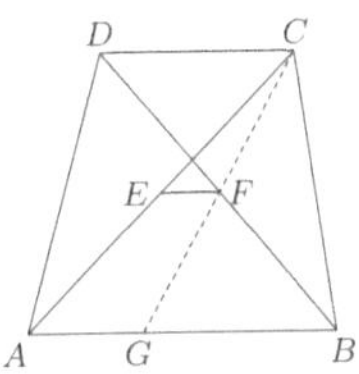

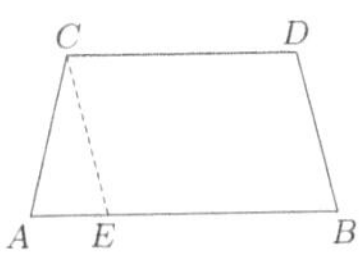

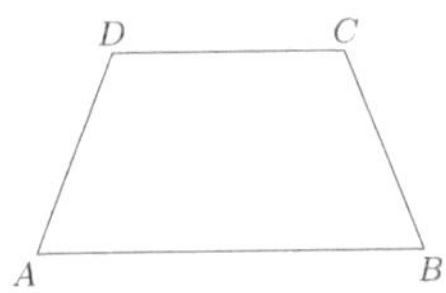

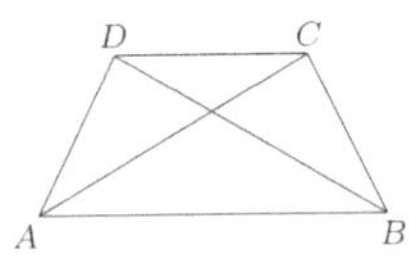

Ex. 84. In an isosceles trapezoid each base makes equal angles with the legs.

Draw $CE \parallel$ to DB. $CE = DB$. (Why?) $\angle A = \angle CEA$, $\angle B = \angle CEA$, $\angle_s C$ and D have equal supplements.

Ex. 85. If the angles at the base of a trapezoid are equal, the other angles are equal, and the trapezoid is isosceles.

Ex. 86. In an isosceles trapezoid the opposite angles are supplementary:

$$\angle C = \angle D \text{ (Ex. 84)}$$

Ex. 87. The diagonals on an isosceles trapezoidal are equal.
Prove $\triangle ACD = \triangle BDC$.

Ex. 88. If the diagonals of a trapezoid are equal, the trapezoid is isosceles.
Draw CE and $DF \perp$ to AB.

$$\triangle ADF = \triangle BCE. \qquad \text{(Why?)}$$

$$\therefore \angle ADF = \angle CBA.$$

$$\triangle ABC = \triangle BAD.$$

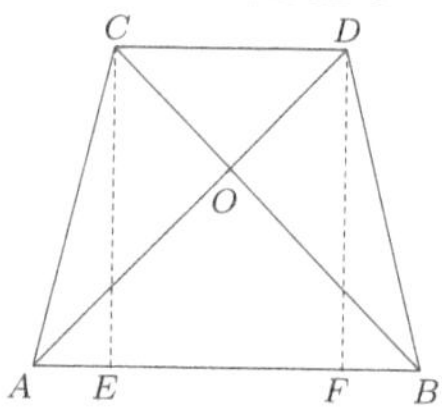

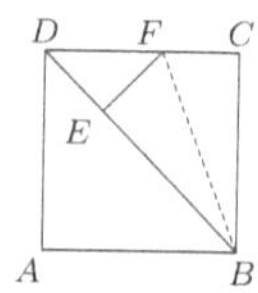

Ex. 89. If from the diagonal DB, of a square $ABCD$, BE is cut off equal to BC, and EF is drawn perpendicular to BD meeting DC at F, then DE is equal to EF and also to FC.

$\angle EDF = 45°$, and $\angle DFE = 45°$; and $DE = DF$. Rt. $\triangle BEF =$ rt. $\triangle BCF$ (§ 151); and $EF = FC$.

Ex. 90. Two angles whose sides are so perpendicular, each to each, are either equal or supplementary.

BOOK II. THE CIRCLE.

DEFINITIONS.

216. A **circle** is a portion of a plane bounded by a curved line, all points of which are equally distant from a point within called the **centre**. The bounding line is called the **circumference** of the circle.

217. A **radius** is a straight line from the centre to the circumference; and a **diameter** is a straight line through the centre, with its ends in the circumference.

By the definition of a circle, *all its radii are equal.* All its diameters are equal, since a diameter is equal to two radii.

218. Postulate. A circumference can be described from any point as a centre, with any given radius.

219. A **secant** is a straight line of unlimited length which intersects the circumference in two points; as, AD (Fig. 1).

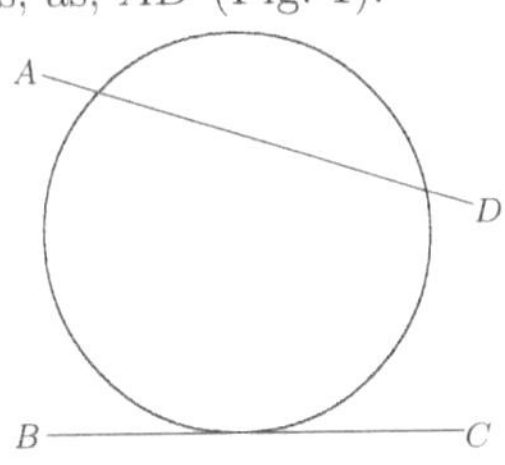

FIG. 1.

220. A **tangent** is a straight line of unlimited length which has one point, and only one, in common with the circumference; as, BC (Fig. 1). In this case the circle is said to be tangent to the straight line. The common point is called the **point of contact**, or **point of tangency**.

221. Two *circles* are tangent to each other, if both are tangent to a straight line at the same point; and are said to be tangent *internally* or *externally*, according as one circle lies wholly *within* or *without* the other.

222. An **arc** is any part of the circumference; as, BC (Fig. 3). Half a circumference is called a **semicircumference**. Two arcs are called **conjugate arcs**, if their sum is a circumference.

223. A **chord** is a straight line that has its extremities in the circumference; as, the straight line BC (Fig. 3).

224. A chord subtends two conjugate arcs. If the arcs are unequal, the less is called the **minor** arc, and the greater the **major** arc. A minor arc is generally called simply an arc.

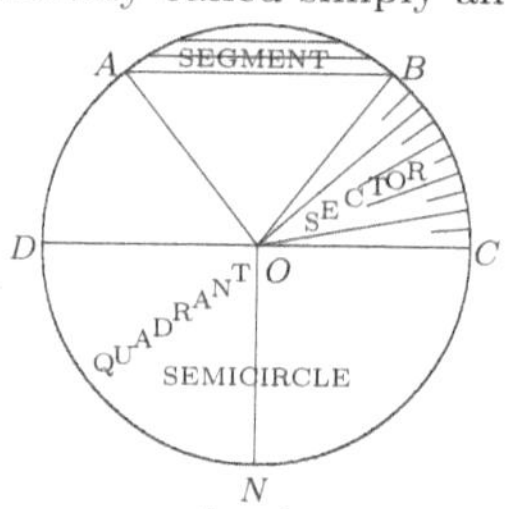

FIG. 2.

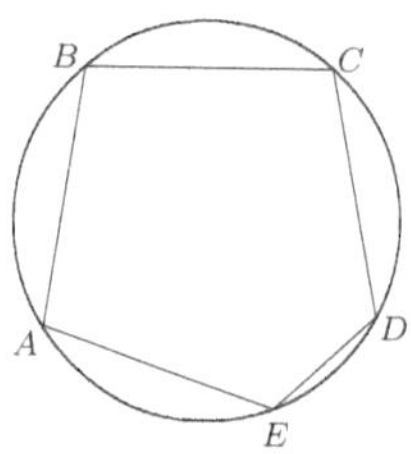

FIG. 3.

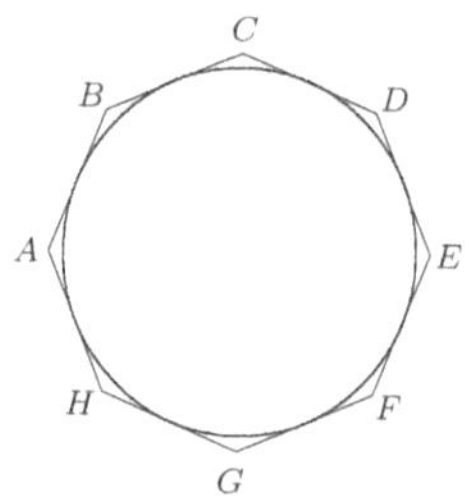

FIG. 4.

225. A **segment** of a circle is a portion of the circle bounded by an arc and its chord (Fig. 2).

226. A **semicircle** is a segment equal to half the circle (Fig. 2).

227. A **sector** of a circle is a portion of the circle bounded by two radii and the arc which they intercept. The angle included by the radii is called the *angle of the sector* (Fig. 2).

228. A **quadrant** is a sector equal to a quarter of the circle (Fig. 2).

229. An angle is called a **central angle**, if its vertex is at the centre and its sides are radii of the circle; as, AOD (Fig. 2).

230. An angle is called an **inscribed angle**, if its vertex is in the circumference and its sides are chords; as, ABC (Fig. 3).

An angle is *inscribed in a segment*, if its vertex is in the arc of the segment and its sides pass through the extremities of the arc.

231. A polygon is *inscribed in a circle*, if its sides are chords; and a circle is *circumscribed about a polygon*, if all the vertices of the polygon are in the circumference (Fig. 3).

232. A circle is *inscribed in a polygon*, if the sides of the polygon are tangent to the circle; and a polygon is *circumscribed about* a circle if its sides are tangents (Fig. 4).

233. *Two circles are equal, if they have equal radii.*

For they will coincide, if their centres are made to coincide.

CONVERSELY: *Two equal circles have equal radii.*

234. *Two circles are concentric*, if they have the same centre.

ARCS, CHORDS, AND TANGENTS.

PROPOSITION I. THEOREM.

235. *A straight line cannot meet the circumference of a circle in more than two points.*

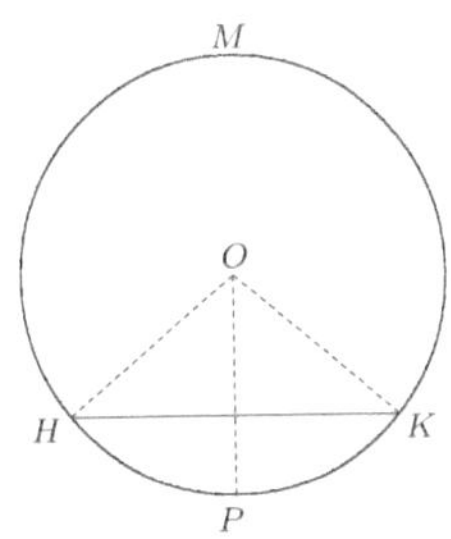

Let HK be any line meeting the circumference HKM in H and K.

To prove that HK cannot meet the circumference in any other point.

Proof. If possible, let HK meet the circumference in P.

Then the radii OH, OP, and OK are equal. § 217

$\therefore$ P does not lie in the straight line HK. § 102

$\therefore$ HK meets the circumference in only two points. Q.E.D.

PROPOSITION II. THEOREM.

236. *In the same circle or in equal circles, equal central angles intercept equal arcs; and of two unequal central angles the greater intercepts the greater arc.*

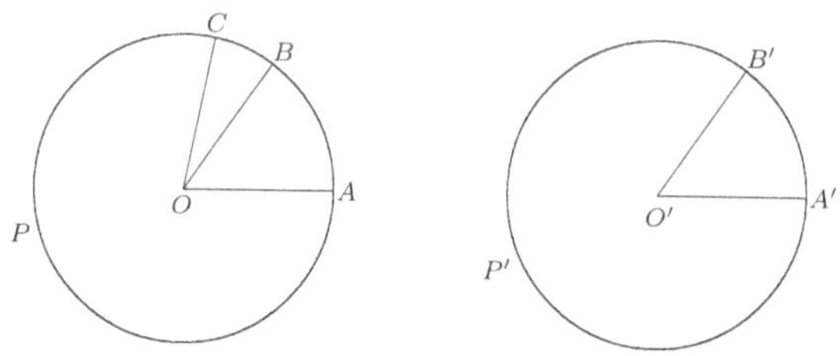

In the equal circles whose centres are O and O', let the angles AOB and $A'O'B'$ be equal, and angle AOC be greater than angle $A'O'C'$.

To prove that 1. arc $AB =$ arc $A'B'$;

2. arc $AC >$ arc $A'B'$.

Proof. 1. Place the $\odot A'B'P'$ on the $\odot ABP$ so that the $A'O'B'$ shall coincide with its equal, the AOB.

Then A' falls on A, and B' on B. § 233

arc $A'B'$ coincides with arc AB. § 216

2. Since the AOC is greater than the $A'O'B'$, it is greater than the AOB, the equal of the $A'O'B'$.

Therefore, OC falls without the AOB.

arc $AC >$ arc AB. Ax. 8

arc $AC >$ arc $A'B'$, the equal of arc AB. Q.E.D.

237. CONVERSELY: *In the same circle or in equal circles, equal arcs subtend equal central angles; and of two unequal arcs the greater subtends the greater central angle.*

To prove that

$$1. \quad AOB = \quad A'O'B';$$

$$2. \quad AOC \text{ is greater than } \quad A'O'B'.$$

Proof. 1. Place the $\odot A'B'P'$ on the $\odot ABP$ so that $O'A$ shall fall on its equal OA, and the arc $A'B'$ on its equal AB.

Then $O'B'$ will coincide with OB. § 47

$A'O'B' = \quad AOB$. § 60

2. Since arc $AC > A'B'$, it is greater than arc AB, the equal of $A'B'$, and OB will fall within the $\quad AOC$.

AOC is greater than $\quad AOB$. Ax. 8

AOC is greater than $\quad A'O'B'$. Q.E.D.

238. COR. 1. *In the same circle or in equal circles, two sectors that have equal angles are equal; two sectors that have unequal angles are unequal, and the greater sector has the greater angle.*

239. COR. 2. *In the same circle or in equal circles, equal sectors have equal angles; and of two unequal sectors the greater has the greater angle.*

240. Law of Converse Theorems. It was stated in § 32 that the converse of a theorem is not necessarily true. If, however, a theorem is in fact a group of three theorems, and if *one of the hypotheses* of the group *must* be true, and *no two of the conclusions can be true at the same time*, then the converse of the theorem is *necessarily* true.

Proposition II. is a group of three theorems. It asserts that the arc AB is equal to the arc $A'B'$, if the angle AOB is equal to the angle $A'O'B'$; that the arc AB is greater than the arc $A'B'$, if the angle AOB is greater than the angle $A'O'B'$; that the arc AB is less than the arc $A'B'$, if the angle AOB is less than the angle $A'O'B'$.

One of these hypotheses must be true; for the angle AOB must be equal to, greater than, or less than, the angle $A'O'B'$.

No two of the conclusions can be true at the same time, for the arc AB cannot be both equal to and greater than the arc $A'B'$; nor can it be both equal to and less than the arc $A'B'$; nor both greater than and less than the arc $A'B'$. In such a case, the converse theorem is *necessarily* true, and no proof like that given in the text is required to establish it.

Proposition III. Theorem.

241. *In the same circle or in equal circles, equal arcs are subtended by equal chords; and of two unequal arcs the greater is subtended by the greater chord.*

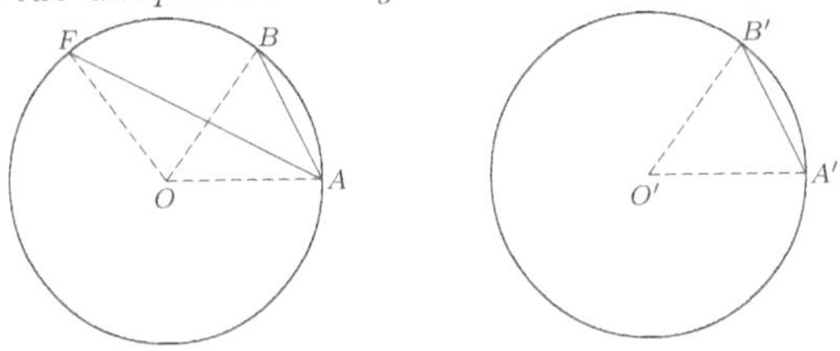

In the equal circles whose centres are O and O', let the arcs AB and $A'B'$ be equal, and the arc AF greater than arc $A'B'$.

To prove that

1. chord $AB =$ chord $A'B'$;

2. chord $AF >$ chord $A'B'$.

Proof.

Draw the radii OA, OB, OF, $O'A'$, $O'B'$.

1.

The ${}_sAOB$ and $A'O'B'$ are equal. § 143

For $OA = O'A'$, and $OB = O'B'$, § 233

(*radii of equal circles*),

and $AOB = A'O'B'$, § 237

(*in equal* $\odot_s$ *equal arcs subtend equal central* ${}_s$).

chord $AB =$ chord $A'B'$. § 128

2.

In the ${}_sAOF$ and $A'O'B'$,

$OA = O'A'$, and $OF = O'B'$. § 233

But the AOF is greater than the $A'O'B'$, § 237

(*in equal* $\odot_s$, *the greater of two unequal arcs subtends the greater*).

chord $AF >$ chord $A'B'$. § 154

Q.E.D.

242. Cor. *In the same circle or in equal circles, the greater of two unequal major arcs is subtended by the less chord.*

Proposition IV. Theorem.

243. Conversely: *In the same circle or in equal circles, equal chords subtend equal arcs; and of two unequal chords the greater subtends the greater arc.*

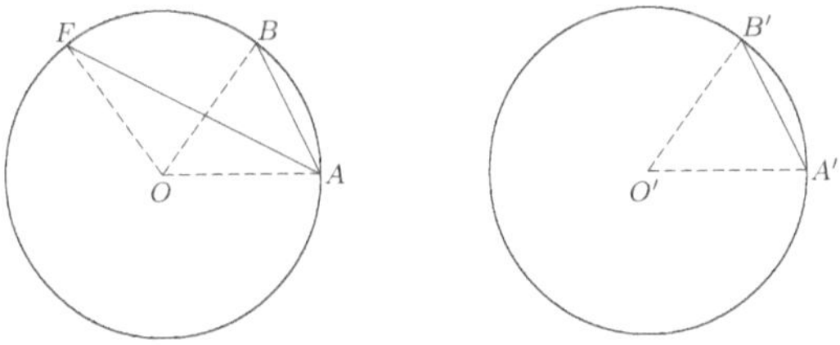

In the equal circles whose centres are O and O', let the chords AB and $A'B'$ be equal, and the chord AF greater than $A'B'$.

To prove that 1. arc AB = arc $A'B'$;

2. arc $AF >$ arc $A'B'$.

Proof.

Draw the radii OA, OB, OF, $O'A'$, $O'B'$.

1.

The $_sOAB$ and $O'A'B'$ are equal. § 150

For $OA = O'A'$, and $OB = O'B'$, § 233

and chord AB = chord $A'B'$. Hyp.

$AOB =$ $A'O'B'$. § 128

arc AB = arc $A'B'$, § 236

(*in equal* $\odot_s$ *equal central* $_s$ *intercept equal arcs*).

2.

In the $_sOAF and O'A'B'$,

$OA = O'A'$ and $OF = O'B'$. § 233

But chord $AF >$ chord $A'B'$. Hyp.

the AOF is greater than the $A'O'B'$. § 155

arc $AF >$ arc $A'B'$, § 236

(*in equal* $\odot_s$ *the greater central* *intercepts the greater arc*).

Q.E.D.

244. Cor. *In the same circle or in equal circles, the greater of two unequal chords subtends the less major arc.*

Proposition V. Theorem.

245. *A diameter perpendicular to a chord bisects the chord and the arcs subtended by it.*

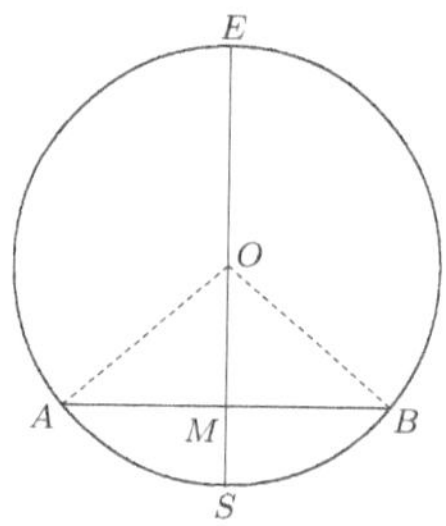

Let ES be a diameter perpendicular to the chord AB at M.

To prove that $AM = BM$, $AS = BS$, and $AE = BE$.

Proof. Draw OA and OB from O, the centre of the circle.

The rt. ${}_sOAM$ and OBM are equal. § 151

For $OM = OM$, Iden.

and $OA = OB$. § 217

$AM = BM$, and $AOS =\quad BOS$. § 128

Likewise $AOE =\quad BOE$. § 85

$AS = BS$, and $AE = BE$. § 236

Q.E.D.

246. Cor. 1. *A diameter bisects the circumference and the circle.*

247. Cor. 2. *A diameter which bisects a chord is perpendicular to it.*

248. Cor. 3. *The perpendicular bisector of a chord passes through the centre of the circle, and bisects the arcs of the chord.*

PROPOSITION VI. THEOREM.

249. *In the same circle or in equal circles, equal chords are equally distant from the centre.* CONVERSELY: *Chords equally distant from the centre are equal.*

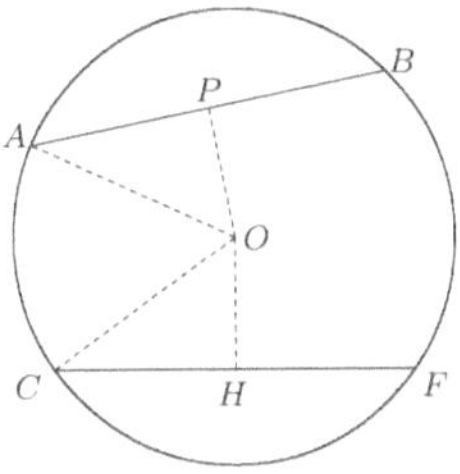

Let AB and CF be equal chords of the circle $ABFC$.

To prove that AB and CF are equidistant from the centre O.

Proof. Draw OP to AB, OH to CF, and join OA and OC.

OP bisects AB, and OH bisects CF. § 245

The rt. $_s OPA$ and OHC are equal. § 151

$$AP = CH,$$ Ax. 7

and

$$OA = OC,$$ § 217

Hence,

$$OP = OH.$$ § 128

AB and CF are equidistant from O.

CONVERSELY:

Let $OP = OH$.

To prove

$$AB = CF.$$

Proof. The rt. $_s OPA$ and OHC are equal. § 151

For

$$OA = OC,$$ § 217

and

$$OP = OH,$$ Hyp.

Hence,

$$AP = CH.$$ § 128

$$AB = CF.$$ Ax. 6

Q.E.D.

Proposition VII. Theorem.

250. *In the same circle or in equal circles, if two chords are unequal, they are unequally distant from the centre; and the greater chord is at the less distance.*

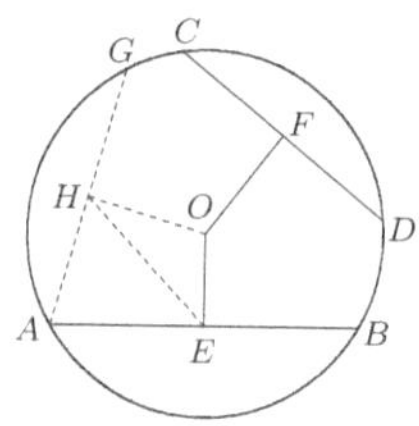

In the circle whose centre is O, let the chords AB and CD be unequal, and AB the greater; and let OE be perpendicular to AB and OF perpendicular to CD.

To prove that

$$OE < OF.$$

Proof. Suppose AG drawn equal to CD, and OH to AG.

Draw EH.

OE bisects AB, and OH bisects AG. § 245

By hypothesis,

$$AB > CD.$$

$AB > AG$, the equal of CD.

$$AE > AH.$$ Ax. 7

AHE is greater than AEH. § 152

OHE, the complement of AHE, is less than OEH, the complement of AEH. Ax. 5

$$OE < OH.$$ § 153

But

$$OH = OF.$$ § 249

$$OE < OF.$$ Q.E.D.

Ex. 91. The perpendicular bisectors of the sides of an inscribed polygon are concurrent (pass through the same point).

PROPOSITION VIII. THEOREM.

251. CONVERSELY: *In the same circle or in equal circles, if two chords are unequally distant from the centre, they are unequal; and the chord at the less distance is the greater.*

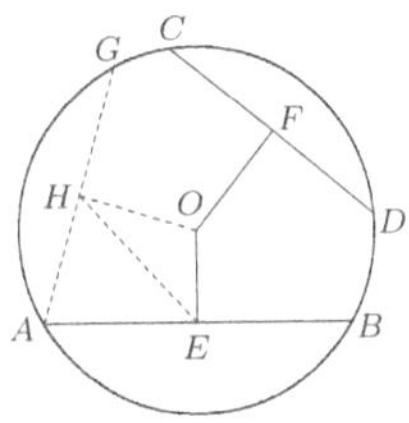

In the circle whose centre is O, let AB and CD be unequally distant from O; and let OE, the perpendicular to AB, be less than OF, the perpendicular to CD.

To prove that

$$AB > CD.$$

Proof. Suppose AG drawn equal to CD, and OH to AG.

Then

$$OH = OF \qquad \text{§ 249}$$

Hence,

$$OE < OH.$$

Draw EH.

OHE is less than OEH. § 152

AHE, the complement of OHE, is greater than AEH, the complement of OEH. Ax. 5

$$AE > AH. \qquad \text{§ 153}$$

But

$$AE = \tfrac{1}{2}AB, \text{ and } AH = \tfrac{1}{2}AG. \qquad \text{§ 245}$$

$$AB > AG. \qquad \text{Ax. 6}$$

But

$$CD = AG. \qquad \text{Const.}$$

$$AB > CD. \qquad \text{Q.E.D.}$$

252. Cor. *A diameter of a circle is greater than any other chord.*

Proposition IX. Theorem.

253. *A straight line perpendicular to a radius at its extremity is a tangent to the circle.*

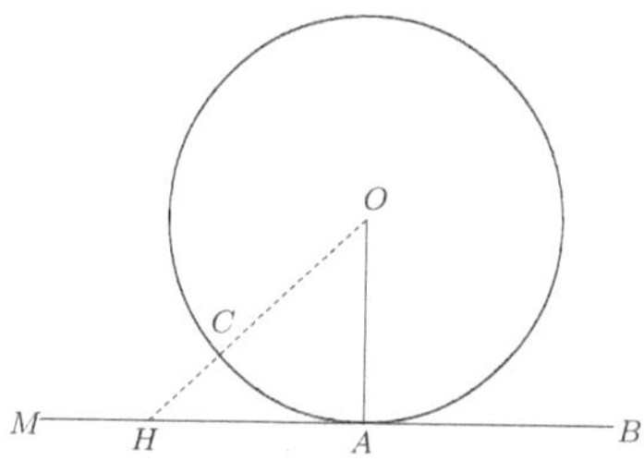

Let MB be perpendicular to the radius OA at A.

To prove that MB is a tangent to the circle.

Proof. From O draw any other line to MB, as OH.

Then

$$OH > OA. \qquad \text{§ 97}$$

the point H is without the circle. § 216

Hence, *every point*, except A, of the line MB is without the circle, and therefore MB is a tangent to the circle at A. § 220

Q.E.D.

254. Cor. 1. *A tangent to a circle is perpendicular to the radius drawn to the point of contact.*

For OA is the shortest line from O to MB, and is therefore   to MB (§ 98); that is, MB is   to OA.

255. Cor. 2. *A perpendicular to a tangent at the point of contact passes through the centre of the circle.*

For a radius is   to a tangent at the point of contact, and therefore a erected at the point of contact coincides with this radius and passes through the centre.

256. Cor. 3. *A perpendicular from the centre of a circle to a tangent passes through the point of contact.*

PROPOSITION X. THEOREM.

257. *Parallels intercept equal arcs on a circumference.*

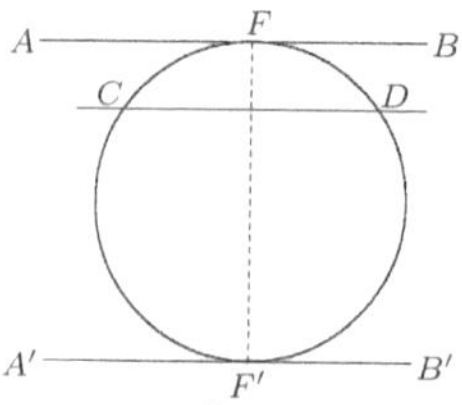

FIG. 1.

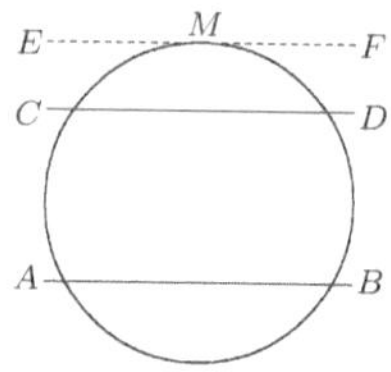

FIG. 2.

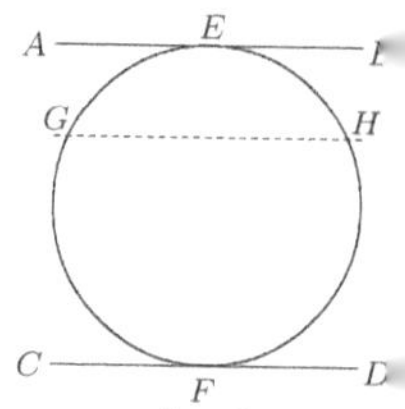

FIG. 3.

CASE 1. **Let AB (Fig. 1) be a tangent at F parallel to CD, a secant.**
To prove that

$$\text{arc}\, CF = \text{arc}\, DF.$$

Proof.

Suppose FF' drawn to AB.

Then FF' is a diameter of the circle. § 255

And FF' is also to CD. § 107

$CF = DF$, and $CF' = DF'$. § 245

CASE 2. **Let** AB **and** CD (Fig. 2) **be parallel secants.**
To prove that

$$\text{arc}\, AC = \text{arc}\, BD.$$

Proof.

Suppose $EF \parallel$ to CD and tangent to the circle at M.

Then

$$\text{arc}\, AM = \text{arc}\, BM,$$ Case 1

and

$$\text{arc}\, CM = \text{arc}\, DM.$$

$$\text{arc}\, AC = \text{arc}\, BD.$$ Ax. 3

CASE 3. **Let** AB **and** CD (Fig. 3) **be parallel tangents at** E **and** F**.**
To prove that

$$\text{arc}\, EGF = \text{arc}\, EHF.$$

Proof.

Suppose GH drawn $\parallel$ to AB.

Then

$$\text{arc } EG = \text{arc } EH, \qquad \text{Case 1}$$

and

$$\text{arc } GF = \text{arc } HF.$$

$$\text{arc } EGF = \text{arc } EHF. \qquad \text{Ax. 2}$$

Q.E.D.

PROPOSITION XI. THEOREM.

258. *Through three points not in a straight line one circumference, and only one, can be drawn.*

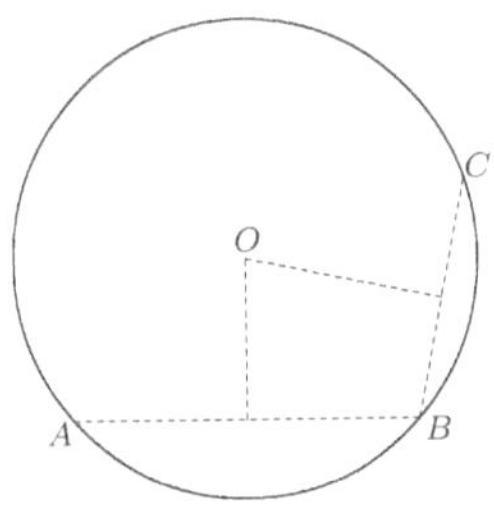

Let A, B, C be three points not in a straight line.

To prove that one circumference, and only one, can be drawn through A, B, and C.

Proof.

Draw AB and BC.

At the middle points of AB and BC suppose ${}_s$ erected.

These ${}_s$ will intersect at some point O, since AB and BC are not in the same straight line.

The point O is in the perpendicular bisector of AB, and is therefore equidistant from A and B; the point O is also in the perpendicular bisector of BC, and is therefore equidistant from B and C. § 160

Therefore, O is equidistant from A, B, and C; and a circumference described from O as a centre, with a radius OA, will pass through the three given points.

The centre of a circumference passing through the three points must be in both perpendiculars, and hence at their intersection. As two straight lines can intersect in only one point, O is the centre of the only circumference that can pass through the three given points. Q.E.D.

259. Cor. *Two circumferences can intersect in only two points.* For, if two circumferences have three points common, they coincide and form one circumference.

260. Def. A **tangent from an external point to a circle** is the part of the tangent between the external point and the point of contact.

Proposition XII. Theorem.

261. *The tangents to a circle drawn from an external point are equal, and make equal angles with the line joining the point to the centre.*

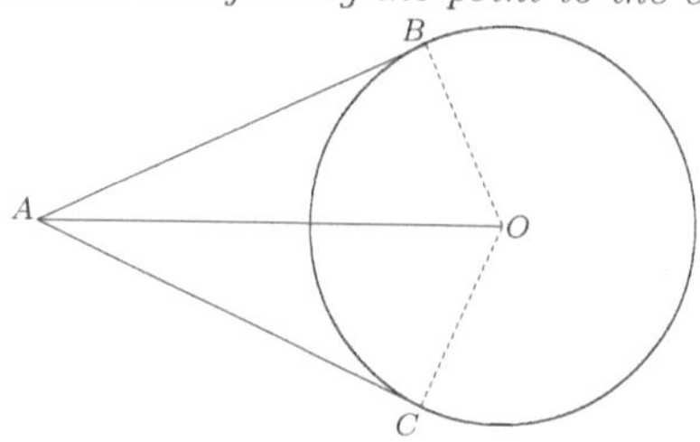

Let AB and AC be tangents from A to the circle whose centre is O, and let AO be the line joining A to the centre O.

To prove that $AB = AC$, *and* $BAO =$ CAO.

Proof.

Draw OB and OC.

AB is to OB, and AC to OC, § 254

(*a tangent to a circle is to the radius drawn to the point of contact*).

The rt. ${}_sOAB$ and OAC are equal. § 151

For OA is common, and the radii OB and OC are equal. § 217

$AB = AC$, and $BAO =$ CAO. § 128

Q.E.D.

262. Def. The line joining the centres of two circles is called the **line of centres**.

263. Def. A tangent to two circles is called a **common external tangent** if it does not cut the line of centres, and a **common internal tangent** if it cuts the line of centres.

PROPOSITION XIII. THEOREM.

264. *If two circles intersect each other, the line of centres is perpendicular to their common chord at its middle point.*

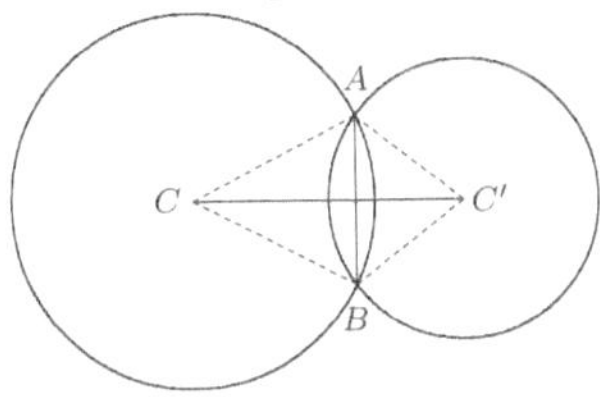

Let C and C' be the centres of the two circles, AB the common chord, and CC' the line of centres.

To prove that CC' is   to AB at its middle point.

Proof.

Draw CA, CB, $C'A$, and $C'B$.

$CA = CB$, and $C'A = C'B$. § 217

C and C' are two points, each equidistant from A and B.

CC' is the perpendicular bisector of AB. § 161

Q.E.D.

Ex. 92. Describe the relative position of two circles if the line of centres:

1. is greater than the sum of the radii;
2. is equal to the sum of the radii;
3. is less than the sum but greater than the difference of the radii;
4. is equal to the difference of the radii;
5. is less than the difference of the radii.

Illustrate each case by a figure.

Ex. 93. The straight line drawn from the middle point of a chord to the middle point of its subtended arc is perpendicular to the chord.

Ex. 94. The line which passes through the middle points of two parallel chords passes through the centre of the circle.

PROPOSITION XIV. THEOREM.

265. *If two circles are tangent to each other, the line of centres passes through the point of contact.*

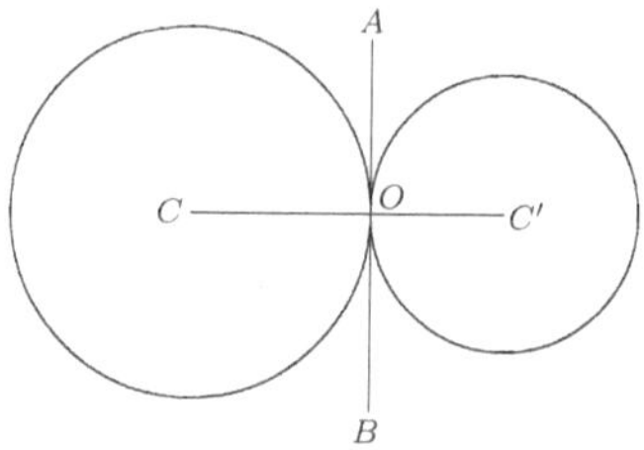

Let the two circles, whose centres are C and C', be tangent to the straight line AB at Q, and CC' the line of centres.

To prove that O is in the straight line CC'.

Proof. A   to AB, drawn through the point O, passes through the centres C and C', § 255

(*a*   *to a tangent at the point of contact passes through the centre of the circle*).

the line CC', having two points in common with this   must coincide with it. § 47

O is in the straight line CC'. Q.E.D.

Ex. 95. Describe the relative position of two circles if they may have:

1. two common external and two common internal tangents;
2. two common external tangents and one common internal tangent;
3. two common external tangents and no common internal tangent;
4. one common external and no common internal tangent;
5. no common tangent.

Illustrate each case by a figure.

Ex. 96. The line drawn from the centre of a circle to the point of intersection of the two tangents is the perpendicular bisector of the chord joining the points of contact.

MEASUREMENT.

266. To **measure** a quantity of any kind is to find *the number of times* it contains a known quantity of the *same kind*, called the **unit of measure**.

The *number* which shows the number of times a quantity contains the unit of measure is called the **numerical measure** of that quantity.

267. No quantity is great or small except by comparison with another quantity of the *same kind*. This comparison is made by finding the numerical measures of the two quantities in terms of a common unit, and then dividing one of the measures by the other.

The quotient is called their **ratio**. In other words the ratio of two quantities of the same kind is the *ratio* of their *numerical measures* expressed in terms of a common unit.

The ratio of a to b is written $a : b$, or $\frac{a}{b}$.

268. Two quantities that can be expressed in *integers* in terms of a common unit are said to be **commensurable**, and the exact value of their ratio can be found. The common unit is called their *common measure*, and each quantity is called a *multiple* of this common measure.

Thus, a common measure of $2\frac{1}{2}$ feet and $3\frac{2}{3}$ feet is $\frac{1}{6}$ of a foot, which is contained 15 times in $2\frac{1}{2}$ feet, and 22 times in $3\frac{2}{3}$ feet. Hence, $2\frac{1}{2}$ feet and $3\frac{2}{3}$ feet are multiples of $\frac{1}{6}$ of a foot, since $2\frac{1}{2}$ feet may be obtained by taking $\frac{1}{6}$ of a foot 15 times, and $3\frac{2}{3}$ feet by taking $\frac{1}{6}$ of a foot 22 times. The ratio of $2\frac{1}{2}$ feet to $3\frac{2}{3}$ feet is expressed by the fraction $\frac{15}{22}$.

269. Two quantities of the same kind that cannot *both* be expressed in *integers* in terms of a common unit, are said to be **incommensurable**, and the *exact value* of their ratio cannot be found. But by taking the unit sufficiently small, an *approximate value* can be found that shall differ from the true value of the ratio by less than any assigned value, however small.

Thus, suppose the ratio, $\frac{a}{b} = \sqrt{2}$.

Now $\sqrt{2} = 1.41421356\cdots$, a value greater than 1.414213, but less than 1.414214.

If, then, a *millionth part* of b is taken as the unit of measure, the value of $\frac{a}{b}$ lies between 1.414213 and 1.414214, and therefore differs from either of these values by less than 0.000001.

By carrying the decimal further, an approximate value may be found that will differ from the true value of the ratio by less than *a billionth, a trillionth, or any other assigned value.*

In general, if $\frac{a}{b} > \frac{m}{n}$ but $< \frac{m+1}{n}$, then the error in taking either of these values for $\frac{a}{b}$ is less than $\frac{1}{n}$, the difference between these two fractions. But by increasing n indefinitely, $\frac{1}{n}$ can be decreased indefinitely, and a value of the ratio can be found within any required degree of accuracy.

270. The ratio of two incommensurable quantities is called an **incommensurable ratio**; and is a *fixed value* which its successive approximate values constantly approach.

THE THEORY OF LIMITS.

271. When a quantity is regarded as having a *fixed* value throughout the same discussion, it is called a **constant**; but when it is regarded, under the conditions imposed upon it, as having *different successive* values, it is called a **variable**.

If a variable, by having different successive values, can be made to differ from a given constant by less than any assigned value, however small, but cannot be made absolutely equal to the constant, that constant is called the **limit** of the variable, and the variable is said to **approach the constant as its limit**.

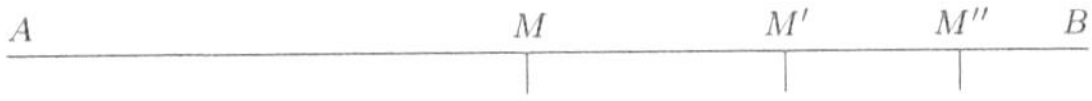

272. Suppose a point to move from A toward B, under the conditions that the first second it shall move one half the distance from A to B, that is, to M; the next second, one half the remaining distance, that is, to M'; and so on indefinitely.

Then it is evident that the moving point *may approach as near to B as we choose, but will never arrive at B.* For, however near it may be to B at any instant, the next second it will pass over half the distance still remaining; it must, therefore, approach nearer to B, since *half* the distance still remaining is *some* distance, but will not reach B, since *half* the distance still remaining is not the *whole* distance.

Hence, the distance from A to the moving point is an increasing variable, which indefinitely approaches the constant AB as its *limit*; and the distance from the moving point to B is a decreasing variable, which indefinitely approaches the *constant zero* as its *limit.*

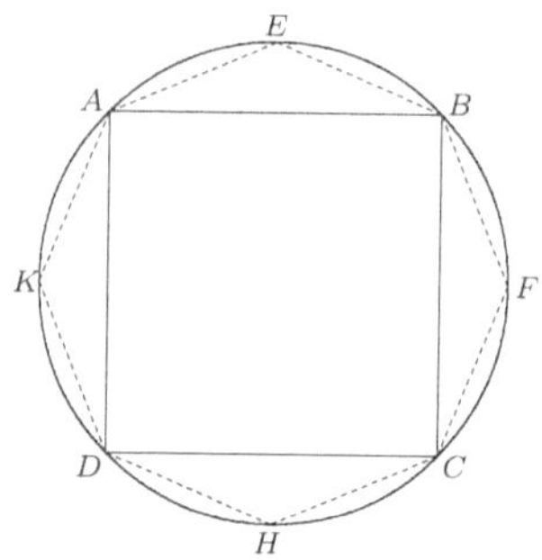

273. Again, suppose a square $ABCD$ inscribed in a circle, and E, F, H, K the middle points of the arcs subtended by the sides of the square. If we draw the lines AE, EB, BF, etc., we shall have an inscribed polygon of double the number of sides of the square.

The length of the perimeter of this polygon, represented by the dotted lines, is greater than that of the square, since two sides replace each side of the square and form with it a triangle, and two sides of a triangle are together greater than the third side; but less than the length of the circumference, for it is made up of straight lines, each one of which is less than the part of the circumference between its extremities.

By continually doubling the number of sides of each resulting inscribed figure, the length of the perimeter will increase with the increase of the number of sides, but will not become equal to the length of the circumference.

The difference between the perimeter of the inscribed polygon and the circumference of the circle can be made less than any assigned value, but cannot be made equal to zero.

The length of the circumference is, therefore, the *limit* of the length of the perimeter as the *number of sides* of the inscribed figure is *indefinitely increased.* § 271

274. Consider the decimal $0.333\cdots$ which may be written

$$\tfrac{3}{10}+\tfrac{3}{100}+\tfrac{3}{1000}+\cdots$$

The value of each fraction after the first is one tenth of the preceding fraction, and by continuing the series we shall reach a fraction less than *any* assigned value, that is, the values of the successive fractions *approach zero as a limit.*

The *sum* of these fractions is less than $\frac{1}{3}$; but the more terms we take, the nearer does the sum *approach* $\frac{1}{3}$ *as a limit.*

275. Test for a limit. In order to prove that a variable approaches a constant as a limit, it is necessary to prove that the difference between the variable and the constant:

1. *Can be made less than any assigned value, however small.*
2. *Cannot be made absolutely equal to zero.*

276. Theorem. *If the limit of a variable x is zero, then the limit of kx, the product of the variable by any finite constant k, is zero.*

1. Let q be any assigned quantity, however small.

Then $\dfrac{q}{k}$ is not 0. Hence x, which may differ as little as we please from 0, may be taken less than $\dfrac{q}{k}$, and then kx will be less than q.

2. Since x cannot be 0, kx cannot be 0.

Therefore, the limit of $kx = 0$ § 275

277. Cor. *If the limit of a variable x is zero, then the limit of the quotient of the variable by any finite constant k, is also zero.*

For $\frac{x}{k} = \frac{1}{k} \times x$, which by § 276 can be made less than any assigned value, however small, but cannot be made equal to zero.

278. Theorem. *The limit of the sum of a finite number of variables x, y, z, $\cdots$ is equal to the sum of their respective limits a, b, c, $\cdots$.*

Let d, d', d'', $\cdots$ denote the differences between x, y, z, $\cdots$ and a, b, c, $\cdots$, respectively. Then $d + d' + d'' + \cdots$ can be made less than any assigned quantity q.

For, if d, d', d'', $\cdots$ are n in number and d is the largest,

$$d + d' + d'' + \cdots < nd. \quad (1)$$

Since d may be diminished at pleasure, we may make d so small that

$$d < \frac{q}{n}; \text{ and therefore } nd < q.$$

But by (1), $d + d' + d'' + \cdots < nd$, and therefore $< q$.

Therefore, the difference between $(x+y+z+\cdots)$ and $(a+b+c+\cdots)$ can be made less than any assigned quantity, but not zero.

Therefore, the limit of $(x + y + z + \cdots) = a + b + c + \cdots$. § 275

279. Theorem. *If the limit of a variable x is not zero, and if k is any finite constant, the limit of the product kx is equal to the limit of x multiplied by k.*

1. If a denotes the limit of x, then x cannot be equal to a. § 271

Therefore, kx cannot be equal to ka.

2. The limit of $(a - x) = 0$. Hence, the limit of $ka - kx = 0$. § 276

Therefore, the limit of $kx = ka$. § 275

280. Cor. *The limit of the quotient of a variable x by any finite constant k is the limit of x divided by k.*

For $\frac{x}{k} = \frac{1}{k} \times x$, and $\frac{\text{the limit of } x}{k} = \frac{1}{k} \times$ the limit of x.

281. Theorem. *The limit of the product of two or more variables is the product of their respective limits, provided no one of these limits is zero.*

If x and y are variables, a and b their respective limits, we may put $x = a - d$, $y = b - d'$; then d and d' are variables which can be made less than any assigned quantity, but not zero. § 275

Now,
$$\begin{aligned} xy &= (a-d)(b-d') \\ &= ab - ad' - bd + dd' \\ ab - xy &= ad' + bd - dd'. \end{aligned}$$

Since every term on the right contains d or d', the whole right member can be made less than any assigned quantity, but not zero. § 278

Hence, $ab - xy$ can be made less than any assigned quantity, but not zero.

Therefore, the limit of $xy = ab$. § 275

Similarly, for three or more variables.

282. Cor. 1. *The limit of the nth power of a variable is the nth power of its limit.*

For the limit of the product of the variables $x, y, z, \cdots$ to n factors is the product of their respective limits, the constants $a, b, c, \cdots$ to n factors (§ 281). If the n factors $xyz \cdots$ are each equal to x, and the n factors $abc \cdots$ are each equal to a, we have $xyz \cdots = x^n$, and $abc \cdots = a^n$.

Therefore, the limit of $x^n = a^n$.

283. Cor. 2. *The limit of the nth root of a variable is the nth root of its limit.*

For if the limit of $x = a$, we may put this in the following form,

$$\text{the limit of } \sqrt[n]{x^n} = \sqrt[n]{a^n};$$

that is, the limit of $\sqrt[n]{xxx \cdots \text{ to } n \text{ factors}}$ is $\sqrt[n]{aaa \cdots \text{ to } n \text{ factors}}$.

Now, $xxx \cdots$ is a variable since each factor is a variable, and $aaa \cdots$ is a constant since each factor is a constant.

If we denote $xxx \cdots$ to n factors by the variable y, and $aaa \cdots$ to n factors by the constant b, we have

$$\text{the limit of } \sqrt[n]{y} = \sqrt[n]{b}.$$

284. Theorem. *If two variables are constantly equal, and each approaches a limit, the limits are equal.*

Let x and y be two variables, a and b their respective limits, d and d' the respective differences between the variables and their limits. Then, if the variables are *increasing* toward their limits

$$a = x + d, \text{ and } b = y + d'.$$

Since the equation $x = y$ is always true, we have by subtraction

$$a - b = d - d'.$$

Since a and b are constants, $a - b$ is a constant; therefore, $d - d'$, which is equal to $a - b$, is a constant.

But the only constant which is less than any assigned value is 0. Therefore, $d - d' = 0$. Therefore, $a - b = 0$, and $a = b$.

If the variables x and y are *decreasing* toward their limits a and b, respectively, then

$$a = x - d \text{ and } b = y - d'.$$

Therefore, by subtraction

$$a - b = d' - d.$$

Therefore, by the same proof as for increasing variables

$$a = b.$$

285. Theorem. *If two variables have a constant ratio, and each approaches a limit that is not zero, the limits have the same ratio.*

Let x and y be two variables, a and b their respective limits.

Let

$$\frac{x}{y} = r, \text{ a constant; then } x = ry.$$

Since x and ry are two variables that are always equal,

the limit of x = the limit of ry. § 284

Now,

the limit of $ry = r \times$ limit of y. § 279

But the limit of x is a, and the limit of y is b.

Therefore,

$$a = rb; \text{ that is, } \frac{a}{b} = r.$$

Proposition XV. Problem.

286. *To find the ratio of two straight lines.*

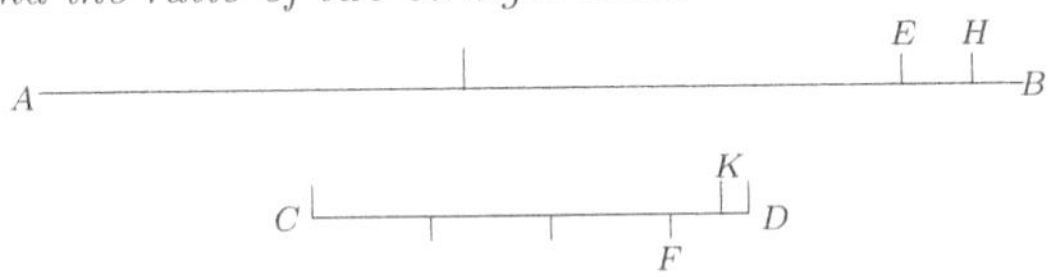

Let AB and CD be two straight lines.

To find the ratio of AB and CD.

Apply CD to AB as many times as possible.

Suppose twice, with a remainder EB.

Then apply EB to CD as many times as possible.

Suppose three times, with a remainder FD.

Then apply FD to EB as many times as possible.

Suppose once, with a remainder HB.

Then apply HB to FD as many times as possible.

Suppose once, with a remainder KD.

Then apply KD to HB as many times as possible.

Suppose KD is contained just twice in HB.

Then

$$HB = 2KD;$$

$$FD = HB + KD = 3KD;$$

$$EB = FD + HB = 5KD;$$

$$CD = 3EB + FD = 18KD;$$

$$AB = 2CD + EB = 41KD;$$

$$\frac{AB}{CD} = \frac{41KD}{18KD} = \frac{41}{18}.$$

Q.E.F.

NOTE. By the same process the ratio of two arcs of the same circle or of equal circles can be found.

If the lines or arcs are incommensurable, an approximate value of the ratio can be found by the same method.

MEASURE OF ANGLES.

PROPOSITION XVI. THEOREM.

287. *In the same circle or in equal circles, two central angles have the same ratio as their intercepted arcs.*

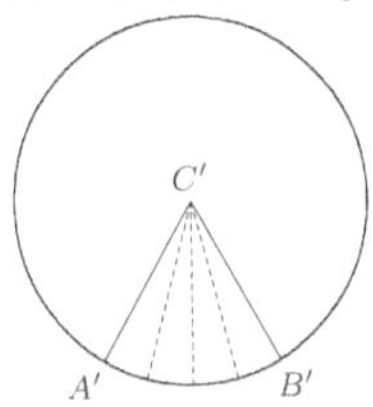

FIG. 1.

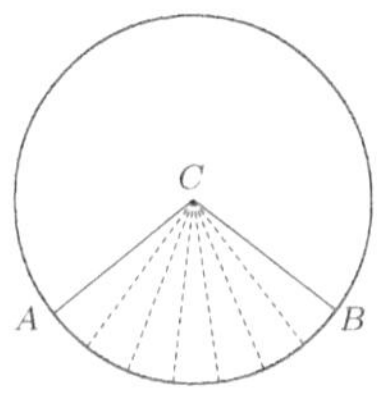

FIG. 2.

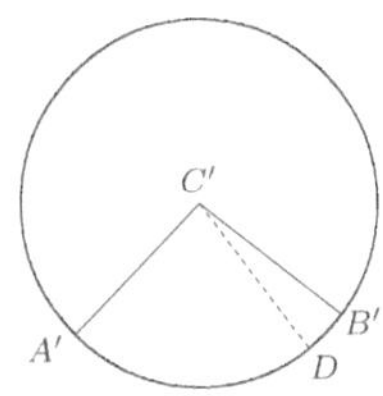

FIG. 3.

In the equal circles whose centres are C and C', let ACB and $A'C'B'$ be the angles, AB and $A'B'$ the intercepted arcs.

To prove that

$$\frac{A'C'B'}{ACB} = \frac{\text{arc } A'B'}{\text{arc } AB}.$$

CASE 1. *When the arcs are commensurable* (Figs. 1 and 2).

Proof. Let the arc m be a common measure of $A'B'$ and AB.

Suppose m to be contained 4 times in $A'B'$,

and 7 times in AB.

Then

$$\frac{\text{arc } A'B'}{\text{arc } AB} = \frac{4}{7}.$$

At the several points of division on AB and $A'B'$ draw radii.

These radii will divide ACB into 7 parts, and $A'C'B'$ into 4 parts, equal each to each, § 237

(*in the same* ⊙, *or equal* $\odot_s$, *equal arcs subtend equal central* $_s$).

$$\frac{A'C'B'}{ACB} = \frac{4}{7}.$$

$$\frac{A'C'B'}{ACB} = \frac{\text{arc } A'B'}{\text{arc } AB}.$$ Ax. 1

CASE 2. *When the arcs are incommensurable* (Figs. 2 and 3).

Proof. Divide AB into any number of equal parts, and apply one of these parts to $A'B'$ as many times as $A'B'$ will contain it.

Since AB and $A'B'$ are incommensurable, a certain number of these parts will extend from A' to some point, as D, leaving a remainder DB' less than one of these parts. Draw $C'D$.

By construction AB and $A'D$ are commensurable.

$$\frac{A'C'D}{ACB} = \frac{\text{arc } A'D}{\text{arc } AB}. \qquad \text{Case 1}$$

By increasing the *number* of equal parts into which AB is divided we can diminish at pleasure the *length* of each part, and therefore make DB' less than any assigned value, however small, since DB' is always less than one of the equal parts into which AB is divided.

We cannot make DB' equal to zero, since, by hypothesis, AB and $A'B'$ are incommensurable. § 269

Hence, DB' approaches zero as a limit, if the number of parts of AB is indefinitely increased. § 275

And the corresponding angle $DC'B'$ approaches zero as a limit.

Therefore, the arc $A'D$ approaches the arc $A'B'$ as a limit, § 271
and the $A'C'D$ approaches the $A'C'B'$ as a limit.

Therefore,

$$\frac{\text{arc } A'D}{\text{arc } AB} \text{ approaches } \frac{\text{arc } A'B'}{\text{arc } AB} \text{ as a limit,} \qquad \text{§ 280}$$

and

$$\frac{A'C'D}{ACB} \text{ approaches } \frac{A'C'B'}{ACB} \text{ as a limit.} \qquad \text{§ 280}$$

But

$$\frac{A'C'D}{ACB} \text{ is constantly equal to } \frac{\text{arc } A'D}{\text{arc } AB},$$

as $A'D$ varies in value and approaches $A'B'$ as a limit.

$$\frac{A'C'B'}{ACB} = \frac{\text{arc } A'B'}{\text{arc } AB}. \qquad \text{§ 284}$$

Q.E.D.

288. A circumference is divided into 360 equal parts, called *degrees*; and therefore a unit angle at the centre intercepts a unit arc on the circumference. Hence, the *numerical measure of a central angle* expressed in terms of the unit angle is equal to the *numerical measure of its intercepted arc* expressed in terms of the unit arc. This must be understood to be the meaning when it is said that

A central angle is measured by its intercepted arc.

PROPOSITION XVII. THEOREM.

289. *An inscribed angle is measured by half the arc intercepted between its sides.*

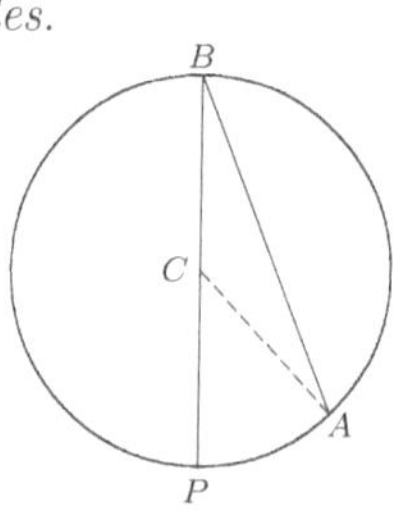

FIG. 1.

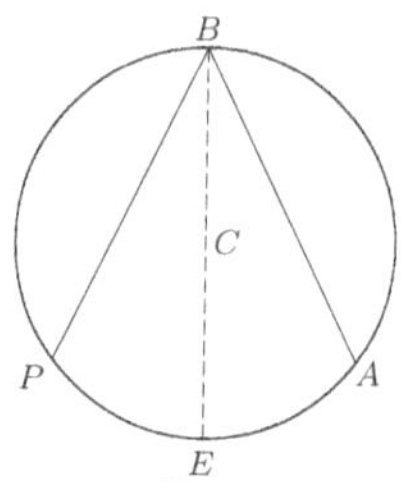

FIG. 2.

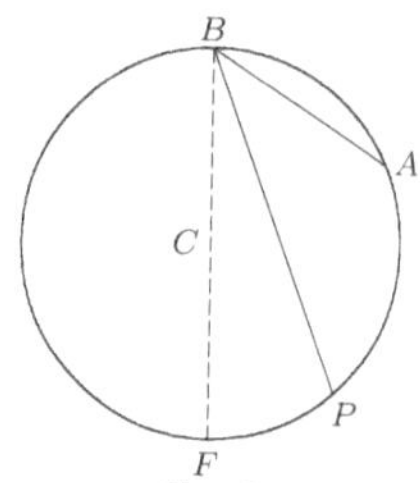

FIG. 3.

1. Let the centre C (Fig. 1) be in one of the sides of the angle.

To prove that the B *is measured by* $\frac{1}{2}$ *the arc* PA.

Proof.

Draw CA.

$$CA = CB. \qquad \S\ 217$$

$$B = \quad A. \qquad \S\ 145$$

But

$$PCA = \quad B + \quad A. \qquad \S\ 137$$

$$PCA = 2 \quad B.$$

But

PCA is measured by arc PA, § 288

(*a central* *is measured by its intercepted arc*).

B is measured by $\frac{1}{2}$ arc PA.

2. Let the centre C (Fig. 2) fall within the angle PBA.

To prove that the PBA *is measured by* $\frac{1}{2}$ *the arc* PA.

Proof.

Draw the diameter BCE.

Then

EBA is measured by $\frac{1}{2}$ arc AE,

and

EBP is measured by $\frac{1}{2}$ arc EP. Case 1

$$EBA + \quad EBP \text{ is measured by } \tfrac{1}{2}(\text{arc } AE + \text{arc } EP),$$

or

$$PBA \text{ is measured by } \tfrac{1}{2} \text{ arc } PA.$$

3. Let the centre C (Fig. 3) **fall without the angle** PBA**.**

To prove that the PBA *is measured by* $\frac{1}{2}$ *the arc* PA.

Proof.

Draw the diameter BCF.

Then

$$FBA \text{ is measured by } \tfrac{1}{2} \text{ arc } FA,$$

and

$$FBP \text{ is measured by } \tfrac{1}{2} \text{ arc } FP. \qquad \text{Case 1}$$

$$FBA - \quad FBP \text{ is measured by } \tfrac{1}{2}(\text{arc } FA - \text{arc } FP),$$

or

$$PBA \text{ is measured by } \tfrac{1}{2} \text{ arc } PA. \qquad \text{Q.E.D.}$$

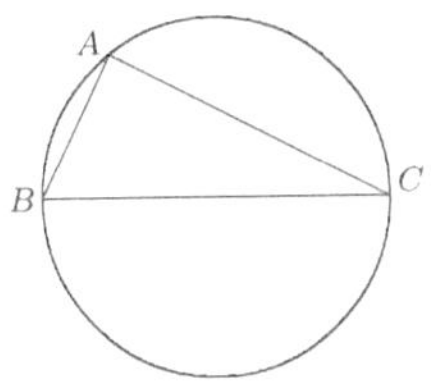

FIG. 4.

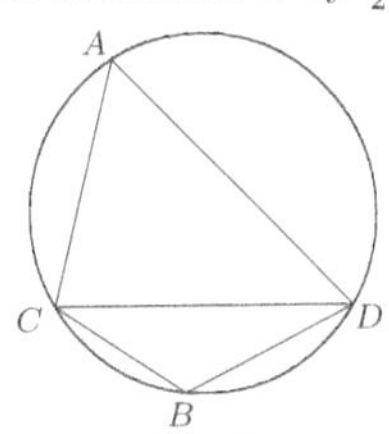

FIG. 5.

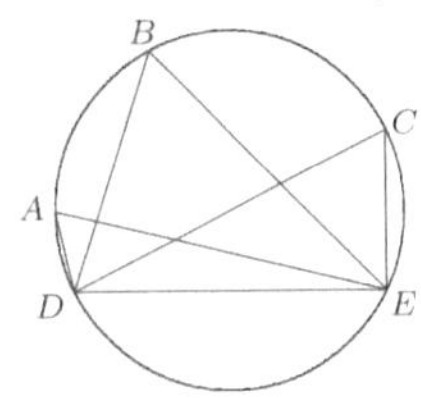

FIG. 6.

290. COR. 1. *An angle inscribed in a semicircle is a right angle.* For it is measured by half a semicircumference (Fig. 4).

291. COR. 2. *An angle inscribed in a segment greater than a semicircle is an acute angle.* For it is measured by an arc less than half a semicircumference; as, CAD (Fig. 5).

292. COR. 3. *An angle inscribed in a segment less than a semicircle is an obtuse angle.* For it is measured by an arc greater than half a semicircumference; as, CBD (Fig. 5).

293. COR. 4. *Angles inscribed in the same segment or in equal segments are equal* (Fig. 6).

Proposition XVIII. Theorem.

294. *An angle formed by two chords intersecting within the circumference is measured by half the sum of the intercepted arcs.*

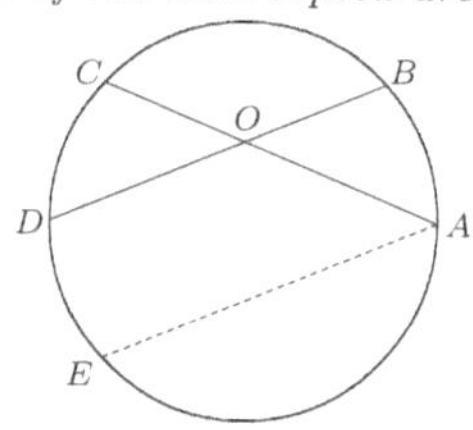

Let the angle COD be formed by the chords AC and BD.

To prove that the COD *is measured by* $\frac{1}{2}(CD + AB)$.

Proof.

Suppose AE drawn $\|$ to BD.

Then arc AB = arc DE, § 257

(*parallels intercept equal arcs on a circumference*).

Also $COD =$ CAE, § 112

(*ext.-int.* $_s$ *of* $\|_s$).

But CAE is measured by $\frac{1}{2}(CD + DE)$, § 289

(*an inscribed* *is measured by half its intercepted arc*).

Put COD for its equal, the CAE, and arc AB for its equal, the arc DE; then COD is measured by $\frac{1}{2}(CD + AB)$.

Q.E.D.

Ex. 97. The opposite angles of an inscribed quadrilateral are supplementary.

Ex. 98. If through a point within a circle two perpendicular chords are drawn, the sum of either pair of the opposite arcs which they intercept is equal to a semicircumference.

Ex. 99. The line joining the centre of the square described upon the hypotenuse of a right triangle to the vertex of the right angle bisects the right angle.

Proposition XIX. Theorem.

295. *An angle included by a tangent and a chord drawn from the point of contact is measured by half the intercepted arc.*

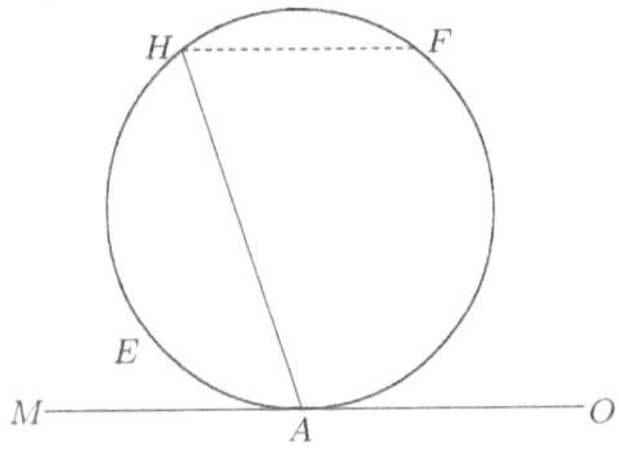

Let MAH be the angle included by the tangent MO to the circle at A and the chord AH.

To prove that the MAH is measured by $\frac{1}{2}$ the arc AEH.

Proof.

Suppose HF drawn $\|$ to MO.

Then arc AF = arc AEH, § 257

(parallels intercept equal arcs on a circumference).

Also $MAH = AHF$, § 110

(alt.-int. $_s$ of $\|_s$).

AHF is measured by $\frac{1}{2}AF$, § 289

(an inscribed is measured by half its intercepted arc).

Put MAH for its equal, the AHF, and arc AEH for its equal, the arc AF; then MAH is measured by $\frac{1}{2}$ arc AEH.

Likewise, the OAH, the supplement of the MAH, is measured by half the arc AFH, the conjugate of the arc AEH.

Q.E.D.

Ex. 100. Two circles are tangent externally at A, and a common external tangent touches them at B and C, respectively. Show that angle BAC is a right angle.

Proposition XX. Theorem.

296. *An angle formed by two secants, two tangents, or a tangent and a secant, drawn to a circle from an external point, is measured by half the difference of the intercepted arcs.*

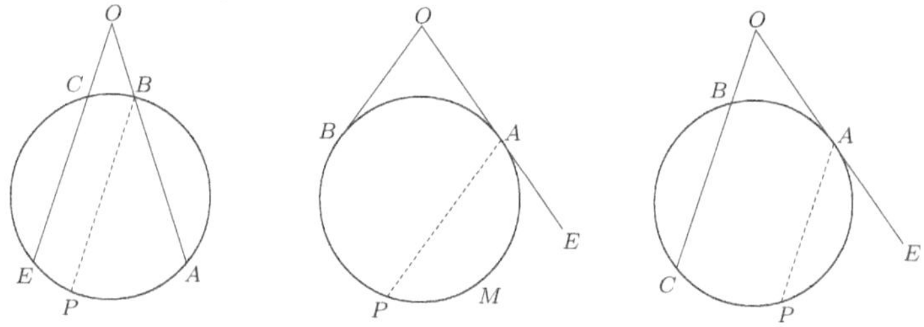

The proof of this theorem is left as an exercise for the student.

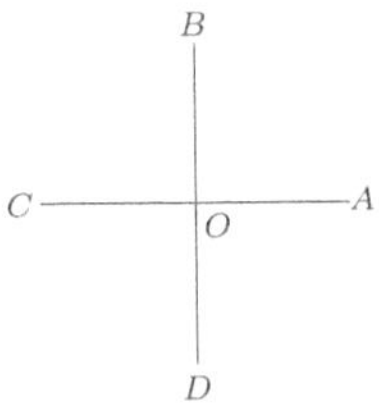

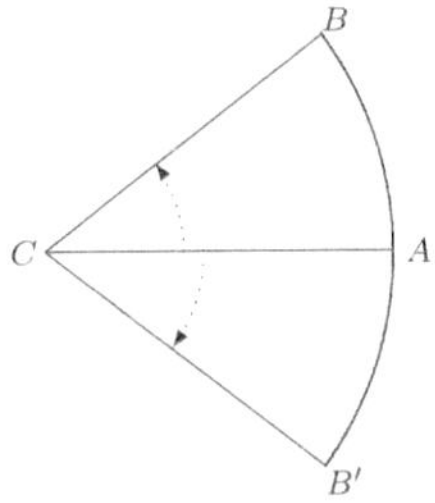

297. Positive and Negative Quantities. In measurements it is convenient to mark the distinction between two quantities that are measured in *opposite directions*, by calling one of them **positive** and the other **negative**.

Thus, if OA is considered positive, then OC may be considered negative, and if OR is considered positive, then OD may be considered negative.

When this distinction is applied to angles, an angle is considered to be *positive*, if the rotating line that describes it moves in the opposite direction to the hands of a clock (counter clockwise), and to be *negative*, if the rotating line moves in the same direction as the hands of a clock (clockwise).

Arcs corresponding to positive angles are considered *positive*, and arcs corresponding to negative angles are considered *negative*.

Thus, the angle ACB described by a line rotating about C from CA to CB is positive, and the arc AB is positive; the angle ACB' described by the line rotating about C from CA to CB' is negative, and the arc AB' is negative.

298. The Principle of Continuity. By marking the distinction between quantities measured in opposite directions, a theorem may often be so stated as to include two or more particular theorems.

The following theorem furnishes a good illustration:

299. *The angle included between two lines of unlimited length that cut or touch a circumference is measured by half the sum of the intercepted arcs.*

Here the word *sum* means the algebraic sum and includes both the arithmetical sum and the arithmetical difference of two quantities.

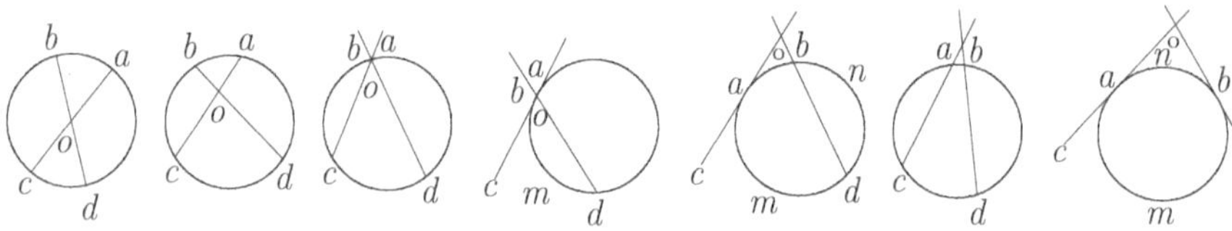

1. If the lines intersect at the centre, the two intercepted arcs are equal, and half the sum will be one of the arcs (§ 288).

2. If the lines intersect between the centre and the circumference, the angle is measured by half the sum of the arcs (§ 294).

3. If the lines intersect on the circumference, one of the arcs becomes zero and we have an inscribed angle (§ 289), or an angle formed by a tangent and a chord (§ 295). In each case the angle is measured by half the sum of the intercepted arcs.

4. If the lines intersect without the circumference, then the arc *ab* is negative and the algebraic sum is the arithmetical difference of the included arcs.

When the reasoning employed to prove a theorem is continued in the manner just illustrated to include two or more theorems, we are said to reason by the *Principle of Continuity*.

REVIEW QUESTIONS ON BOOK II.

1. What do we call the locus of points in a plane that are equidistant from a fixed point in the plane?
2. What does the chord of a segment become when the segment is a semicircle?
3. What kind of an angle do the radii of a sector include when the sector is a semicircle?
4. What is the difference between a chord and a secant?
5. What part of a tangent is meant by a tangent to a circle from an external point?
6. Two chords are equal in equal circles under either of two conditions. What are the two conditions?
7. Points that lie in a straight line are called *collinear*; points that lie in a circumference are called *concyclic*. How many collinear points can be concyclic?
8. What is meant by the statement that a central angle is measured by the arc intercepted between its sides?
9. What is an inscribed angle? What is its measure?
10. What kind of an angle is the angle inscribed in a semicircle? in a segment less than a semicircle? in a segment greater than a semicircle?
11. What is the measure of an angle included by two intersecting chords? by two secants intersecting without the circle?
12. What is the measure of an angle included by a tangent and a chord drawn to the point of contact?
13. When are two quantities of the same kind incommensurable?
14. When are two quantities of the same kind commensurable?
15. Define a variable and the limit of a variable.
16. Does the series $\frac{1}{2}$ in., $\frac{3}{4}$ in., $\frac{7}{8}$ in., $\frac{15}{16}$ in., etc., constitute a variable? Is the variable increasing or decreasing?
17. What is the limit of this variable?
18. What is the test of a limit?

THEOREMS.

Ex. 101. An angle formed by a tangent and a chord is equal to the angle inscribed in the opposite segment.

Ex. 102. Two chords drawn perpendicular to a third chord at its extremities are equal.

Ex. 103. The sum of two opposite sides of a circumscribed quadrilateral is equal to the sum of the other two sides.

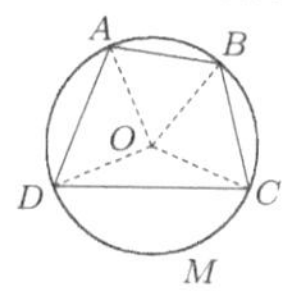

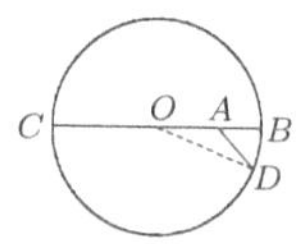

Ex. 104. If the sum of two opposite angles of a quadrilateral is equal to two right angles, a circle may be circumscribed about the quadrilateral.

Let $A+$ $C = 2$ rt. $_s$. Pass a circumference through D, A, and B, and prove that this circumference passes through C.

Ex. 105. The shortest line that can be drawn from a point within a circle to the circumference is the shorter segment of the diameter through that point.

Let A be the given point. Prove AB shorter than any other line AD from A to the circumference.

Ex. 106. The longest line that can be drawn from a point within a circle to the circumference is the longer segment of the diameter through that point.

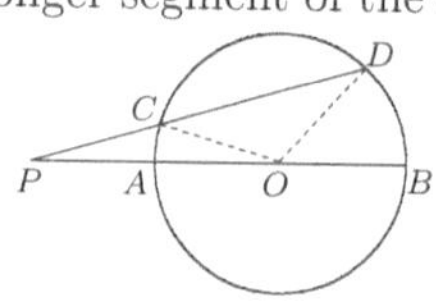

Ex. 107. The shortest line that can be drawn from a point without a circle to the circumference will pass through the centre of the circle if produced.

Ex. 108. The longest line that can be drawn from a point without a circle to the concave arc of the circumference passes through the centre of the circle.

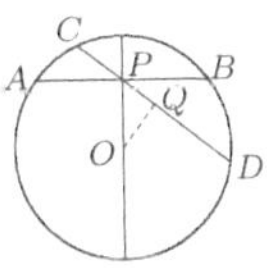

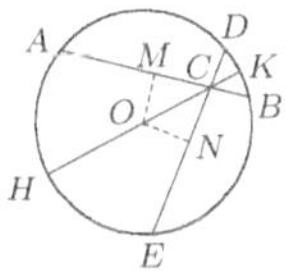

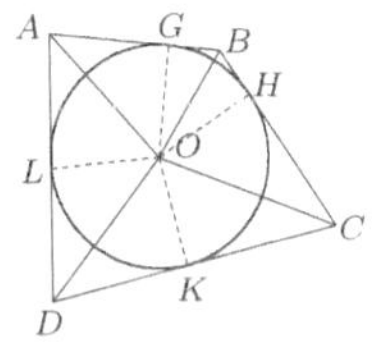

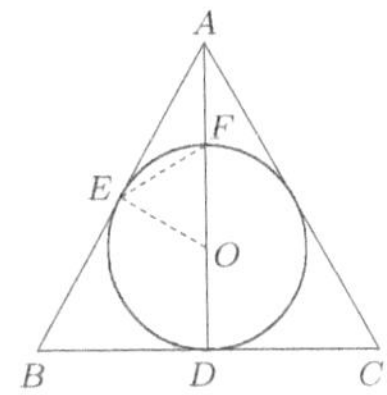

Ex. 109. The shortest chord that can be drawn through a point within a circle is perpendicular to the diameter at that point.

Ex. 110. If two intersecting chords make equal angles with the diameter drawn through the point of intersection, the two chords are equal.

$$\text{Rt. } COM = \text{rt. } CON. \qquad OM = ON.$$

Ex. 111. The angles subtended at the centre of a circle by any two opposite sides of a circumscribed quadrilateral are supplementary.

Ex. 112. The radius of a circle inscribed in an equilateral triangle is equal to one third the altitude of the triangle.

OEF is equiangular and equilateral; $FEA = FAE$.

$$AF = EF. \qquad AF = FO = OD.$$

Ex. 113. The radius of a circle circumscribed about an equilateral triangle is equal to two thirds the altitude of the triangle (Ex. 27).

Ex. 114. A parallelogram inscribed in a circle is a rectangle.

Ex. 115. A trapezoid inscribed in a circle is an isosceles trapezoid.

Ex. 116. All chords of a circle which touch an interior concentric circle are equal, and are bisected at the point of contact.

Ex. 117. If the inscribed and circumscribed circles of a triangle are concentric, the triangle is equilateral (Ex. 116).

Ex. 118. If two circles are tangent to each other the tangents to them from any point of the common internal tangent are equal.

Ex. 119. If two circles touch each other and a line is drawn through the point of contact terminated by the circumferences, the tangents at its ends are parallel.

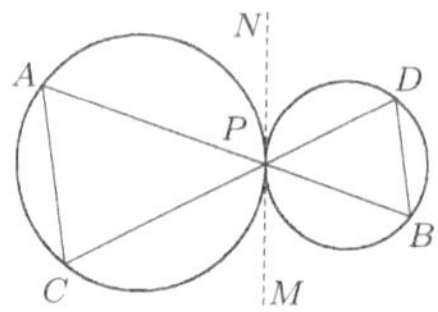

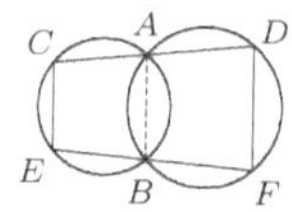

Ex. 120. If two circles touch each other and two lines are drawn through the point of contact terminated by the circumferences, the chords joining the ends of these lines are parallel.

$A = \quad MPC$ and $B = \quad NPD.$ $\qquad A = \quad B.$

Ex. 121. If two circles intersect and a line is drawn through each point of intersection terminated by the circumferences, the chords joining the ends of these lines are parallel.

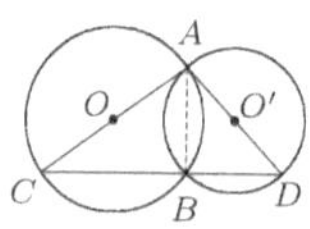

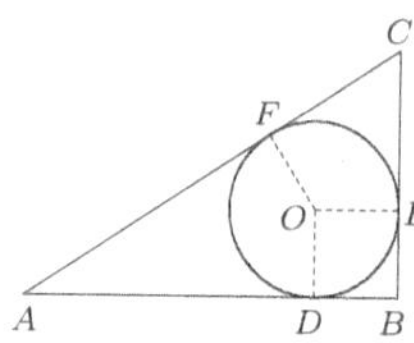

Ex. 122. Through one of the points of intersection of two circles a diameter of each circle is drawn. Prove that the line joining the ends of the diameters passes through the other point of intersection.

$$ABC = \quad ABD = 90^\circ \qquad \S\ 290$$

Ex. 123. If two common external tangents or two common internal tangents are drawn to two circles, the segments intercepted between the points of contact are equal.

Ex. 124. The diameter of the circle inscribed in a right triangle is equal to the difference between the sum of the legs and the hypotenuse.

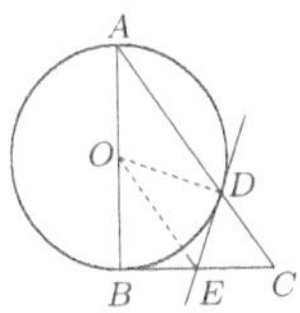

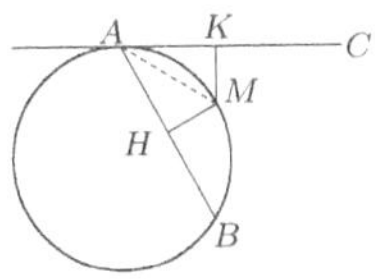

Ex. 125. If one leg of a right triangle is the diameter of a circle, the tangent at the point where the circumference cuts the hypotenuse bisects the other leg.

$$BOE = DOE. \qquad BOE = OAD.$$
$$OE \text{ and } AC \text{ are } \parallel. \qquad BE = EC \ (\S\ 188).$$

Ex. 126. If, from any point in the circumference of a circle, a chord and a tangent are drawn, the perpendiculars dropped on them from the middle point of the subtended arc are equal. $BAM = CAM$.

Ex. 127. The median of a trapezoid circumscribed about a circle is equal to one fourth the perimeter of the trapezoid (Ex. 103).

Ex. 128. Two fixed circles touch each other externally and a circle of variable radius touches both externally. Show that the difference of the distances from the centre of the variable circle to the centres of the fixed circles is constant.

Ex. 129. If two fixed circles intersect, and circles are drawn to touch both, show that either the sum or the difference of the distances of their centres from the centres of the fixed circles is constant, according as they touch (i) one internally and one externally, (ii) both internally or both externally.

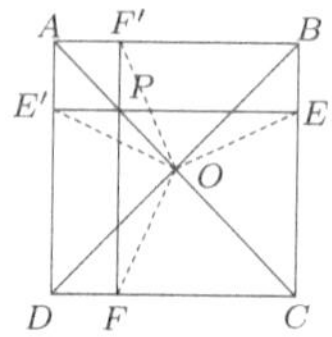

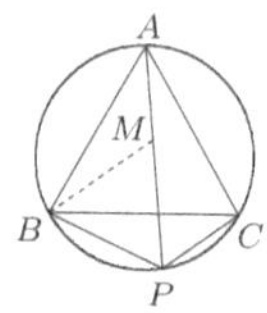

Ex. 130. If two straight lines are drawn through any point in a diagonal of a square parallel to the sides of the square, the points where these lines meet the sides lie on the circumference of a circle whose centre is the point of intersection of the diagonals.

$$POE = \quad POF \text{ (§ 143)}. \qquad OE = OF. \qquad POE' = \quad POF'.$$

Ex. 131. If ABC is an inscribed equilateral triangle and P is any point in the arc BC, then $PA = PB + PC$.

Take $PM = PB$. $ABM = \quad CBP$ (§ 143) and $AM = PC$.

Ex. 132. The tangents drawn through the vertices of an inscribed rectangle, which is not a square, enclose a rhombus.

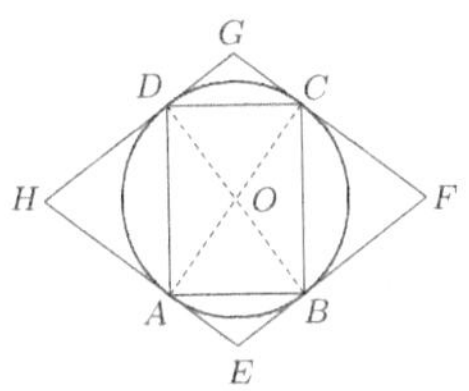

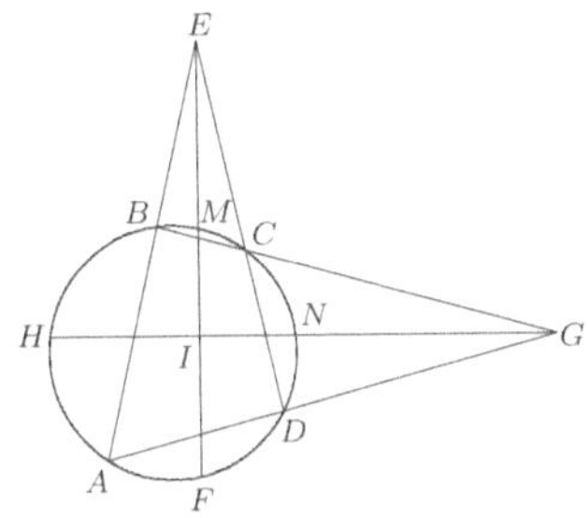

Ex. 133. The bisectors of the angles included by the opposite sides (produced) of an inscribed quadrilateral intersect at right angles.

$$\text{Arc } AF - \text{arc } BM = \text{arc } DF - \text{arc } CM$$

$$\text{and arc } AH - \text{arc } DN = \text{arc } BH - \text{arc } CN.$$

$$\text{arc } FH + \text{arc } MN = \text{arc } HM + \text{arc } FN.$$

$$FIH = \quad HIM.$$

Discussion. This problem is impossible, if any two sides of the quadrilateral are parallel.

PROBLEMS OF CONSTRUCTION.

NOTE. Hitherto we have supposed the figures constructed. We now proceed to explain the methods of constructing simple problems, and afterwards to apply these methods to the solution of more difficult problem.

Proposition XXI. Problem.

300. *To let fall a perpendicular upon a given line from a given external point.*

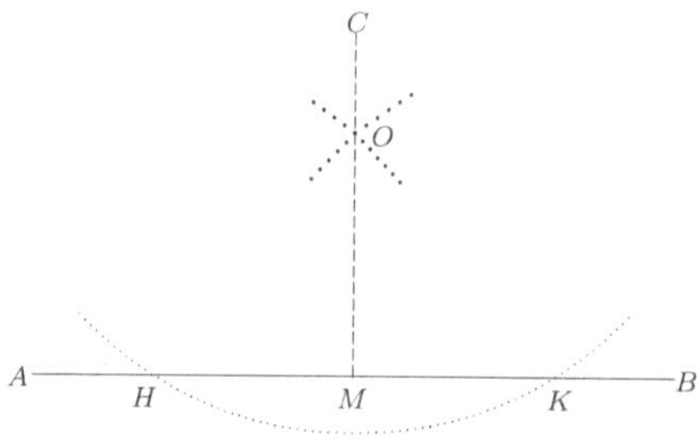

Let AB be the given straight line, and C the given external point.

To let fall a   to the line AB from the point C.

From C as a centre, with a radius sufficiently great, describe an arc cutting AB in two points, H and K.

From H and K as centres, with equal radii greater than $\frac{1}{2}HK$,

describe two arcs intersecting at O.

Draw CO,

and produce it to meet AB at M.

CM is the   required.

Proof. Since C and O are two points each equidistant from H and K, they determine a   to HK at its middle point. § 161

Q.E.F.

Note. *Given* lines of the figures are represented by full lines, *resulting* lines by long-dashed, and *auxiliary* lines by short-dashed lines.

PROPOSITION XXII. PROBLEM.

301. *At a given point in a straight line, to erect a perpendicular to that line.*

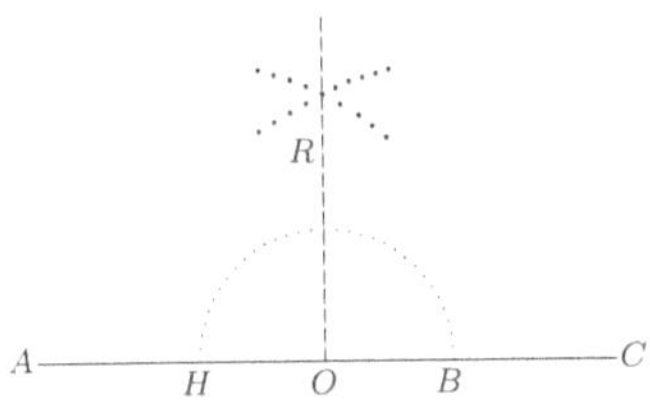

FIG. 1.

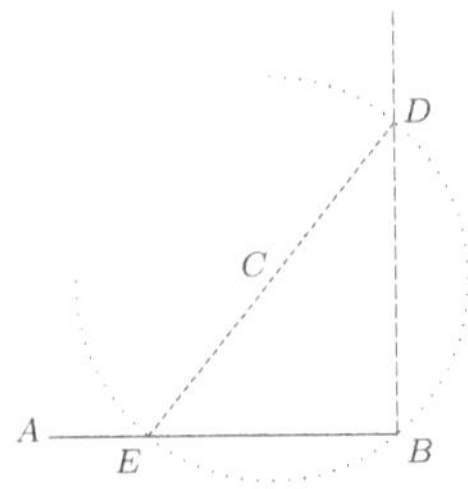

FIG. 2.

1. Let O be the given point in AC. Fig. 1.

Take OH and OB equal.

From H and B as centres, with equal radii greater than OB, describe two arcs intersecting at R. Join OR.

Then the line OR is the required.

Proof. O and R, two points each equidistant from H and B, determine the perpendicular bisector of HB. § 161

2. Let B be the given point. Fig. 2.

Take any point C without AB; and from C as a centre, with the distance CB as a radius, describe an arc intersecting AB at E.

Draw EC, and prolong it to meet the arc again at D.

Join BD, and BD is the required.

Proof.

The B is a right angle. § 290

BD is to AB. Q.E.F.

Discussion. The point C must be so taken that it will not be in the required perpendicular.

Proposition XXIII. Problem.

302. *To bisect a given straight line.*

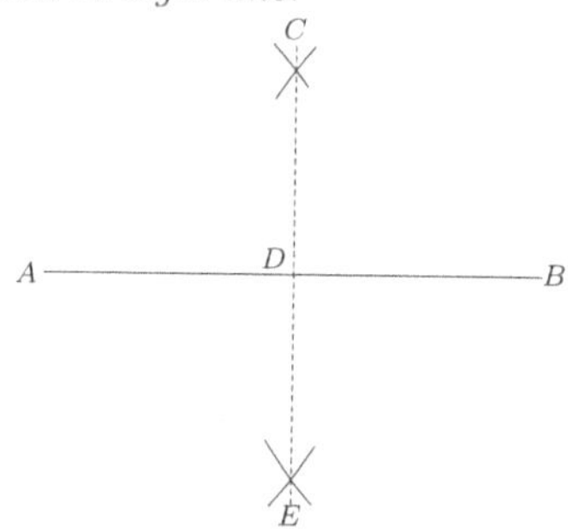

To bisect the given straight line AB.

From A and B as centres, with equal radii greater than $\frac{1}{2}AB$, describe arcs intersecting at C and E.

Join CE.

Then CE bisects AB. § 161

Q.E.F.

Proposition XXIV. Problem.

303. *To bisect a given arc.*

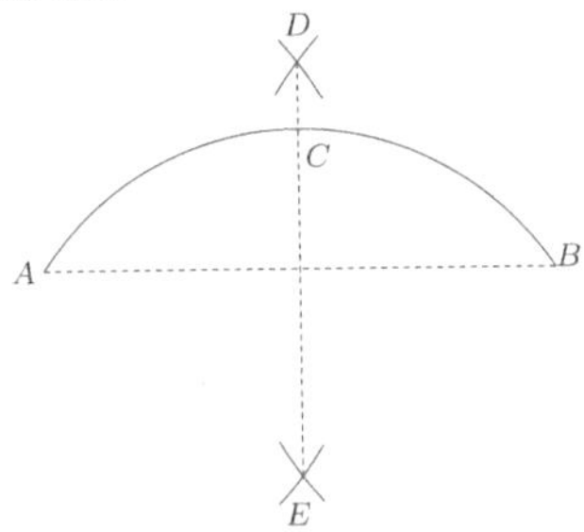

To bisect the given arc *AB*.

Draw the chord AB.

From A and B as centres, with equal radii greater than $\frac{1}{2}AB$, describe arcs intersecting at D and E.

Draw DE.

Then DE is the bisector of the chord AB. § 161

DE bisects the arc ACB. § 248

Q.E.F.

Proposition XXV. Problem.

304. *To bisect a given angle.*

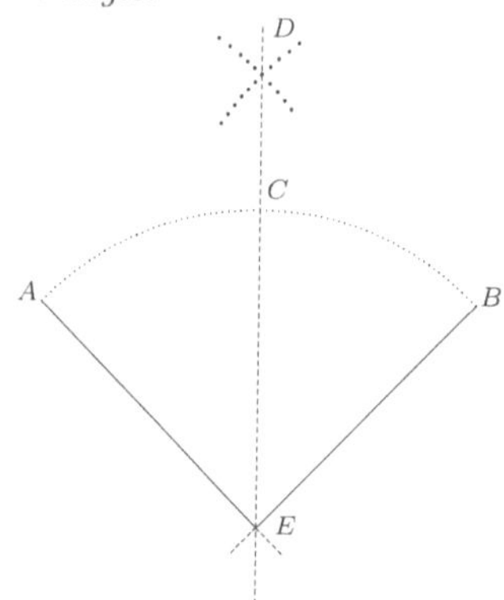

Let AEB be the given angle.

From E as a centre, with any radius, as EA, describe an arc cutting the sides of the E at A and B.

From A and B as centres, with equal radii greater than half the distance from A to B, describe two arcs intersecting at D.

Draw DE.

Then DE bisects the arc AB at C. § 303

DE bisects the angle E. § 237

Q.E.F.

Ex. 134. To construct an angle of 45°; of 135°.

Ex. 135. To construct an equilateral triangle, having given one side.

Ex. 136. To construct an angle of 60°; of 150°.

Ex. 137. To trisect a right angle.

PROPOSITION XXVI. PROBLEM.

305. *At a given point in a given straight line, to construct an angle equal to a given angle.*

At C in the line CM, construct an angle equal to the given angle A.

From A as a centre, with any radius, AE, describe an arc cutting the sides of the A at E and F.

From C as a centre, with a radius equal to AE,

describe an arc HG cutting CM at H.

From H as a centre, with a radius equal to the chord EF,

describe an arc intersecting the arc HG at O.

Draw CO, and HCO is the required angle. Why?

Q.E.F.

Proposition XXVII. Problem.

306. *To draw a straight line parallel to a given straight line through a given external point.*

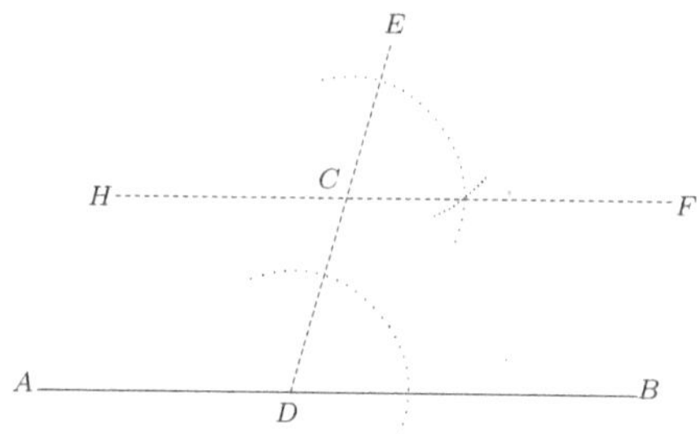

Let AB be the given line, and C the given point.

Draw ECD, making any convenient EDB.

At the point C construct ECF equal to EDB. § 305

Then the line HCF is $\parallel$ to AB. Why?

Q.E.F.

Proposition XXVIII. Problem.

307. *To divide a given straight line into a given number of equal parts.*

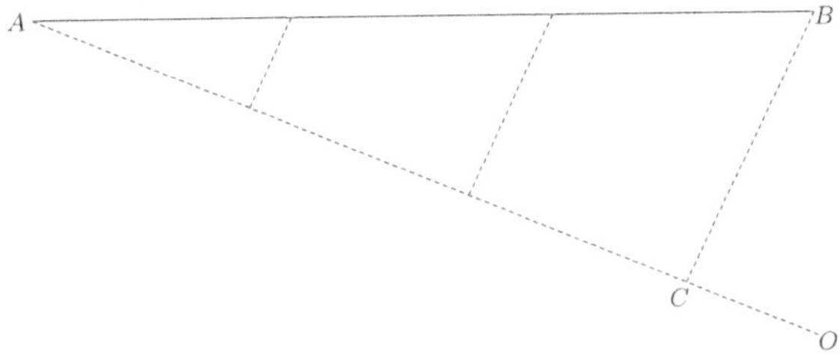

Let AB be the given straight line.

From A draw the line AO, making any convenient angle with AB.

Take any convenient length, and apply it to AO as many times as the line AB is to be divided into parts.

From C, the last point thus found on AO, draw CB.

Through the points of division on AO draw parallels to the line CB. § 306

These lines will divide AB into equal parts. § 187

Q.E.F.

Ex. 138. To construct an equilateral triangle, having given the perimeter.

Ex. 139. To divide a line into four equal parts by two different methods.

Ex. 140. Through a given point to draw a line which shall make equal angles with the two sides of a given angle.

Through the given point draw a   to the bisector of the given .

Ex. 141. To draw a line through a given point, so that it shall form with the sides of a given angle an isosceles triangle (Ex. 140).

Proposition XXIX. Problem.

308. *To find the third angle of a triangle when two of the angles are given.*

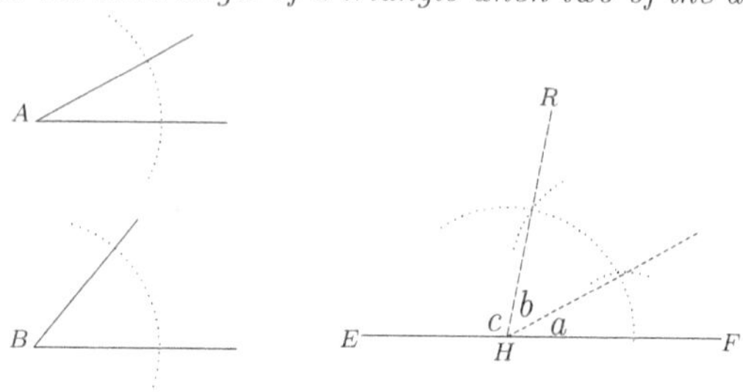

Let A and B be the two given angles.

At any point H in any line EF,

construct a equal to A, and b equal to B. § 305

Then

c is the required. Why?

Q.E.F.

PROPOSITION XXX. PROBLEM.

309. *To construct a triangle when two sides and the included angle are given.*

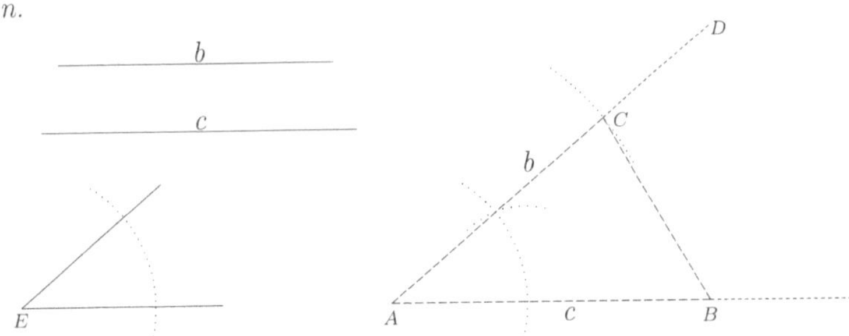

Let b and c be the two sides of the triangle and E the included angle.

Take AB equal to the side c.

At A, construct BAD equal to the given E. § 305

On AD take AC equal to b, and draw CB.

Then ACB is the required. Q.E.F.

PROPOSITION XXXI. PROBLEM.

310. *To construct a triangle when a side and two angles of the triangle are given.*

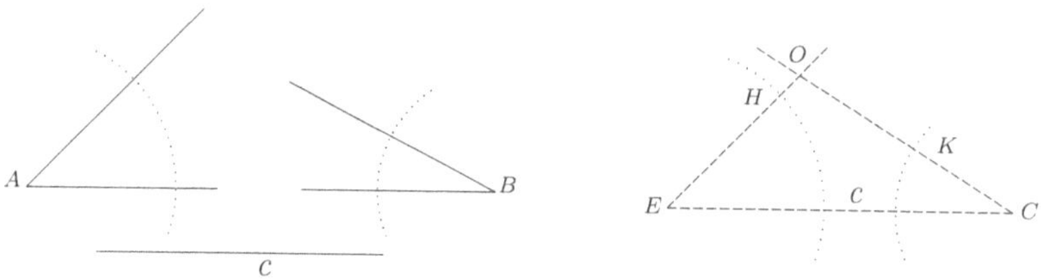

Let c be the given side, A and B the given angles.

Take EC equal to the side c.

At E construct the CEH equal to A. § 305

At C construct the ECK equal to B.

Produce EH and CK until they intersect at O.

Then COE is the required. Q.E.F.

REMARK. If one of the given angles is opposite to the given side, find the third angle by § 308, and proceed as above.

Discussion. The problem is impossible when the two given angles are together equal to or greater than two right angles.

Ex. 142. To construct an equilateral triangle, having given the altitude.

To construct an isosceles triangle, having given:

Ex. 143. The base and the altitude.

Ex. 144. The altitude and one of the legs.

Ex. 145. The angle at the vertex and the altitude.

PROPOSITION XXXII. PROBLEM.

311. *To construct a triangle when two sides and the angle opposite one of them are given.*

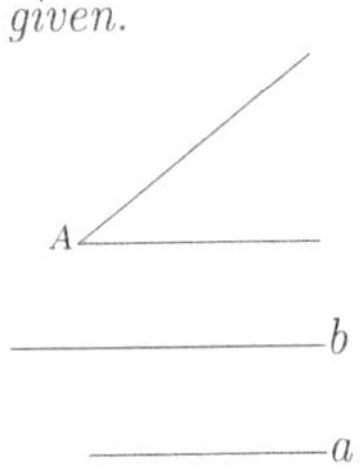

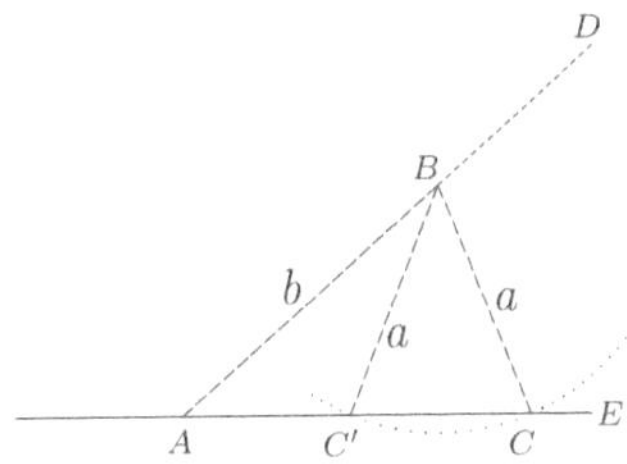

Let a and b be the given sides, and A the angle opposite a.

CASE 1. *If a is less than b.*

Construct DAE equal to the given A § 305

On AD take AB equal to b.

From B as a centre, with a radius equal to a,

describe an arc intersecting the line AE at C and C'.

Draw BC and BC'.

Then both the ${}_{s}ABC$ and ABC' fulfil the conditions, and hence we have two constructions.

This is called the *ambiguous* case.

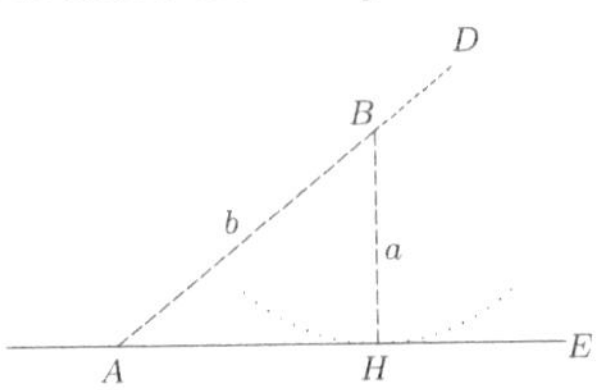

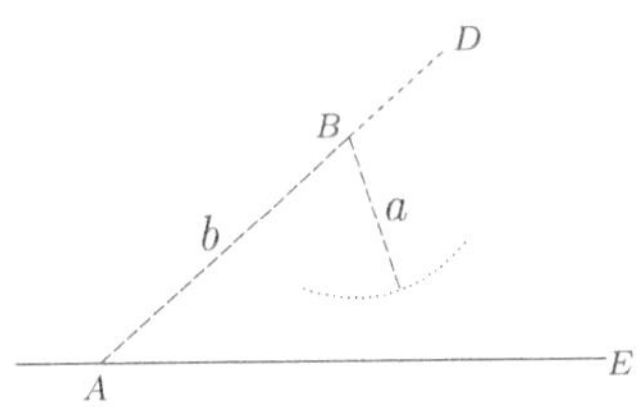

Discussion. If the side a is equal to the BH, the arc described from B will touch AE, and there will be but one construction, the right ABH.

If the given side a is less than the from B, the arc described from B will not intersect or touch AE, and hence the problem is impossible.

If the A is right or obtuse, the problem is impossible; for the side opposite a right or obtuse angle is the greatest side. § 153

CASE 2. *If a is equal to b.*

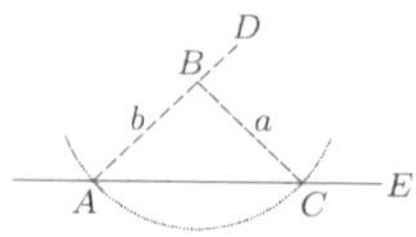

If the A is acute, and $a = b$, the arc described from B as a centre, and with a radius equal to a, will cut the line AE at the points A and C. There is therefore but one solution: the isosceles ABC.

Discussion. If the A is right or obtuse, the problem is impossible; for equal sides of a have equal $_s$ opposite them, and a cannot have two right $_s$ or two obtuse $_s$.

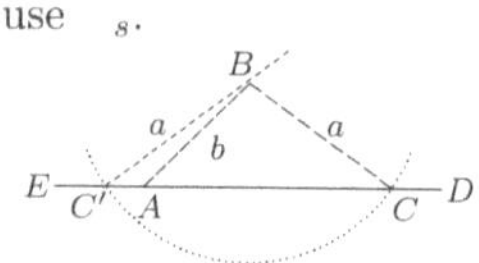

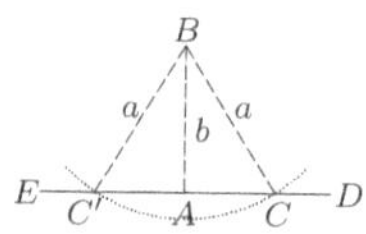

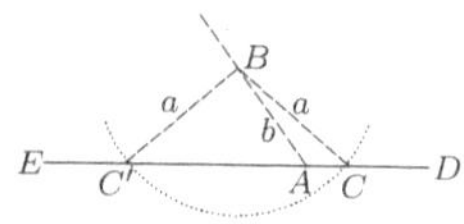

CASE 3. *If a is greater than b.*

If the given A is acute, the arc described from B will cut the line ED on opposite sides of A, at C and C'. The ABC answers the required conditions, but the ABC' does not, for it does not contain the acute A. There is then only one solution; namely, the ABC.

If the A is right, the arc described from B cuts the line ED on opposite sides of A, and we have two *equal* right $_s$ which fulfil the required conditions.

If the A is obtuse, the arc described from B cuts the line ED on opposite sides of A, at the points C and C'. The ABC answers the required conditions, but the ABC' does not, for it does not contain the obtuse A. There is then only one solution; namely, the ABC.

Q.E.F.

PROPOSITION XXXIII. PROBLEM.

312. *To construct a triangle when the three sides of the triangle are given.*

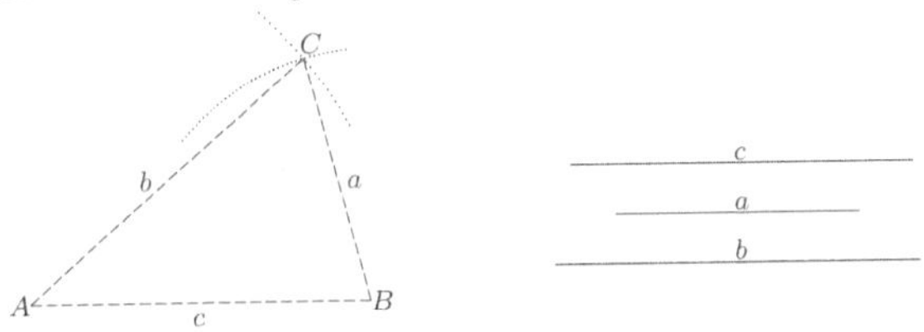

Let the three sides be c, a, and b.

Take AB equal to c. From A as a centre, with a radius equal to b, describe an arc. From B as a centre, with a radius equal to a, describe an arc, intersecting the other arc at C.

Draw CA and CB.

CAB is the required. Q.E.F.

Discussion. The problem is impossible when one side is equal to or greater than the sum of the other two sides.

Proposition XXXIV. Problem.

313. *To construct a parallelogram when two sides and the included angle are given.*

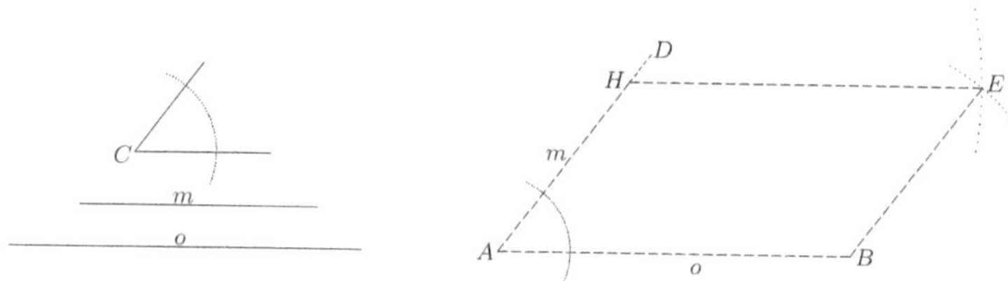

Let m and o be the two sides, and C the included angle.

Take AB equal to o.

At A construct BAD equal to C. § 305

Take AH equal to m. From H as a centre, with a radius equal to o, describe an arc, and from B as a centre, with a radius equal to m, describe an arc, intersecting the other arc at E; and draw EH and EB.

The quadrilateral $ABEH$ is the ▱ required. § 182

Q.E.F.

PROPOSITION XXXV. PROBLEM.

314. *To circumscribe a circle about a given triangle.*

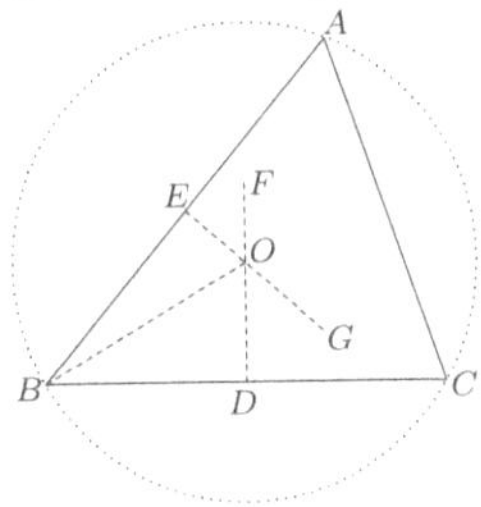

Let ABC be the given triangle.

Bisect AB and BC. § 302

At E and D, the points of bisection, erect $_s$. § 301

Since BC is not the prolongation of AB, these $_s$ will intersect at some point O.

From O, with a radius equal to OB, describe a circle.

The ⊙ABC is the ⊙ required.

Proof.

The point O is equidistant from A and B,

and also is equidistant from B and C. § 160

the point O is equidistant from A, B, and C,

and a ⊙ described from O as a centre, with a radius equal to OB, will pass through the vertices A, B, and C. Q.E.F.

The same construction serves to describe a circumference which shall pass through three points not in the same straight line; also to find the centre of a given circle or of a given arc.

NOTE. The point O is called the *circum-centre* of the triangle.

PROPOSITION XXXVI. PROBLEM.

315. *To inscribe a circle in a given triangle.*

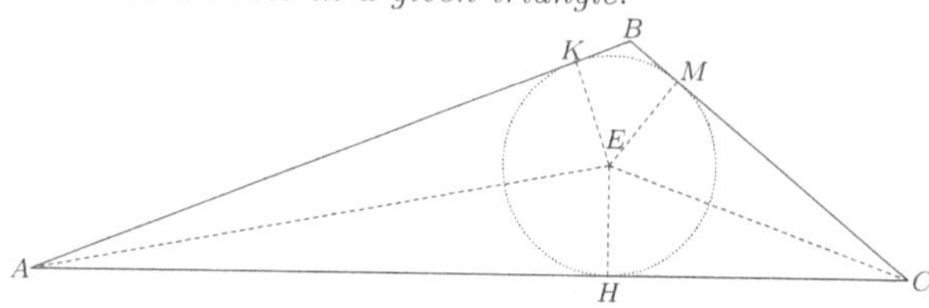

Let ABC be the given triangle.

Bisect the $_s A$ and C. § 304

From E, the intersection of the bisectors,

draw EH to the side AC. § 300

From E as centre, with radius EH, describe the $\odot KHM$.

The $\odot KHM$ is the $\odot$ required.

Proof. Since E is in the bisector of the A, it is equidistant from the sides AB and AC; and since E is in the bisector of the C, it is equidistant from the sides AC and BC. § 162

a $\odot$ described from E as centre, with a radius equal to EH, will touch the sides of the and be inscribed in it. Q.E.F.

NOTE. The point E is called the *in-centre* of the triangle.

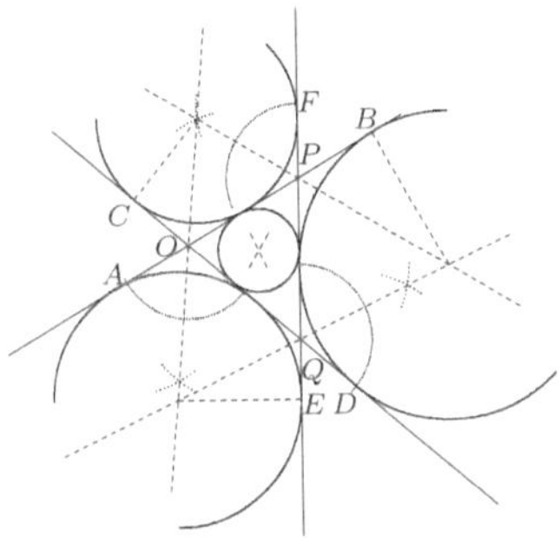

316. The intersections of the bisectors of the exterior angles of a triangle are the centres of three circles, each of which will touch one side of the triangle, and the two other sides produced. These three circles are called *escribed* circles; and their centres are called the *ex-centres* of the triangle.

PROPOSITION XXXVII. PROBLEM.

317. *Through a given point, to draw a tangent to a given circle.*

CASE 1. *When the given point is on the circumference.*

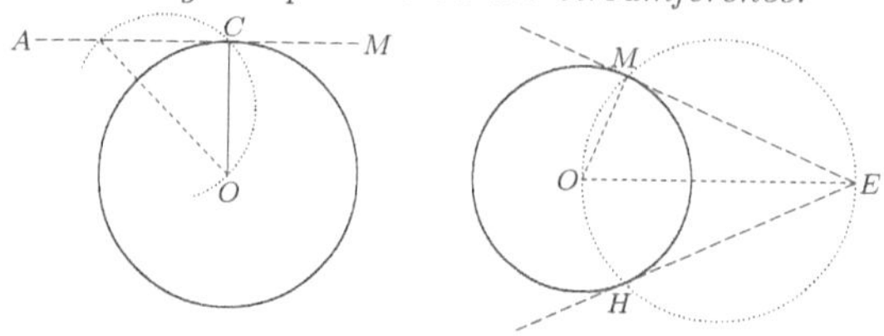

Let C be the given point on the circumference whose centre is O.

From the centre O draw the radius OC.

Through C draw AM to OC. § 301

Then AM is the tangent required. § 253

CASE 2. *When the given point is without the circle.*

Let O be the centre of the given circle, E the given point.

Draw OE.

On OE as a diameter, describe a circumference intersecting the given circumference at the points M and H.

Draw OM and EM.

Then EM is the tangent required.

Proof.

OME is a right angle. § 290

EM is tangent to the circle at M. § 253

In like manner, we may prove EH tangent to the given ⊙.

Q.E.F.

Ex. 146. To draw a tangent to a given circle, so that it shall be parallel to a given straight line.

PROPOSITION XXXVIII. PROBLEM.

318. *Upon a given straight line, to describe a segment of a circle in which a given angle may be inscribed.*

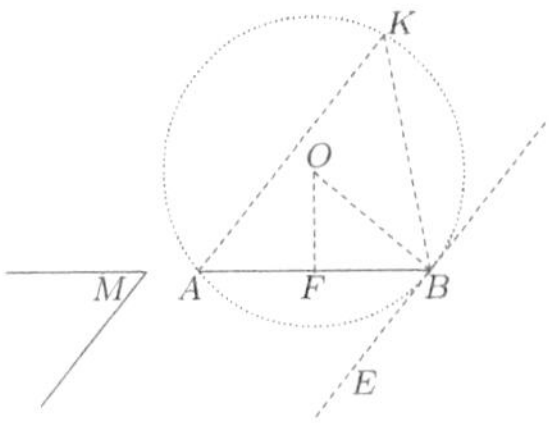

Let AB be the given line, and M the given angle.

Construct the ABE equal to the M. § 305

Bisect the line AB by the OF. § 302

From the point B draw BO to EB. § 301

From O, the point of intersection of FO and BO, as a centre with a radius equal to OB, describe a circumference.

The segment AKB is the segment required.

Proof.

The point O is equidistant from A and B. § 160

 the circumference will pass through A.

But BE is to OB. Const.

 BE is tangent to the ⊙, § 253

(*a straight line to a radius at its extremity is tangent to the* ⊙).

 ABE is measured by $\frac{1}{2}$ arc AB, § 295

(*being an formed by a tangent and a chord*).

But any as K inscribed in the segment AKB is measured by $\frac{1}{2}$ arc AB. § 289

 the M may be inscribed in the segment AKB.

Q.E.F.

SOLUTION OF PROBLEMS.

319. If a problem is so simple that the solution is obvious from a known theorem, we have only to make the construction according to the theorem, and then give a synthetic proof, if a proof is necessary, that the construction is correct, as in the examples of the fundamental problems already given.

320. But problems are usually of a more difficult type. The application of known theorems to their solution is not immediate, and often far from obvious. To discover the mode of application is the first and most difficult part of the solution. The best way to attack such problems is by a method resembling the analytic proof of a theorem, called the **analysis** of the problem.

1. **Suppose the construction made**, and let the figure represent all parts concerned, both given and required.

2. Study the relations among the parts with the aid of known theorems, and try to find some relation that will suggest the construction.

3. If this attempt fails, introduce new relations by drawing auxiliary lines, and study the new relations. If this attempt fails, make a new trial, and so on till a clue to the right construction is found.

321. A problem is *determinate* if it has a *definite* number of solutions, *indeterminate* if it has an *indefinite* number of solutions, and *impossible* if it has *no* solution. A problem is sometimes determinate for certain relative positions or magnitudes of the given parts, and indeterminate for other positions or magnitudes of the given parts.

322. The **discussion** of a problem consists in examining the problem with reference to all possible conditions, and in determining the conditions necessary for its solution.

Ex. 147. PROBLEM. *To construct a circle that shall pass through a given point and cut chords of a given length from two parallels.*

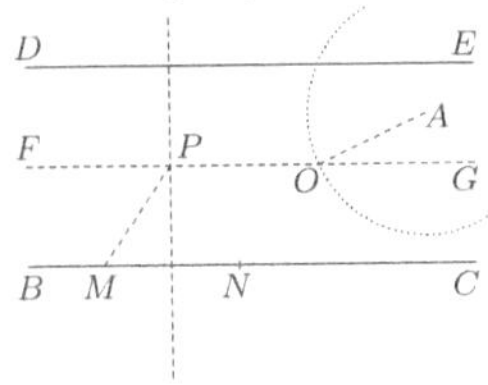

Analysis. Suppose the problem solved. Let A be the given point, BC and DE the given parallels, MN the given length, and O the centre of the required circle.

Since the circle cuts equal chords from two parallels its centre must be equidistant from them. Therefore, one locus for O is $FG \parallel$ to BC and equidistant from BC and DE.

Draw the bisector of MN, cutting FG in P. PM is the radius of the circle required. With A as centre and radius PM describe an arc cutting FG at O. Then O is the centre of the required circle.

Discussion. The problem is impossible if the distance from A to FG is greater than PM.

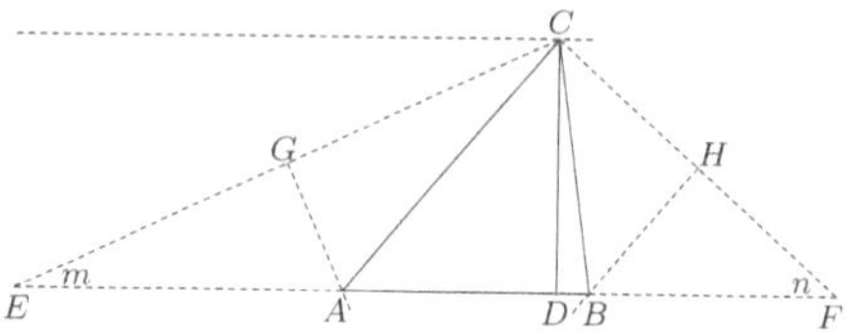

Ex. 148. PROBLEM. *To construct a triangle, having given the perimeter, one angle, and the altitude from the vertex of the given angle.*

Analysis. Suppose the problem solved, and let ABC be the required, ACB the given , and CD the given altitude.

Produce AB both ways, and take $AE = AC$, and $BF = BC$, then $EF =$ the given perimeter. Join CE and CF, forming the isosceles $_sCAE$ and CBF.

In the ECF, $E+$ $F+$ $ECF = 180°$ (why?), but $ECF =$ $ECA+$ $FCB+$ ACB.

Since $E =$ ECA and $F =$ FCB, we have $ECF =$ $E+$ $F+$ ACB. 2 $E + 2$ $F+$ $ACB = 180°$.

$E+$ $F + \frac{1}{2}$ $ACB = 90°$, and $E+$ $F = 90° - \frac{1}{2}$ ACB.

By substitution, $ECF = 90° + \frac{1}{2}$ ACB.

ECF is known.

Construction. To find the point C, construct on EF a segment that will contain the ECF (§ 318), and draw a parallel to EF at the distance CD, the given altitude.

To find the points A and B, draw the bisectors of the lines CE and CF, and the points A and B will be vertices of the required . Why?

PROBLEMS OF CONSTRUCTION.

Ex. 149. Find the locus of a point at a given distance from a given circumference.

Find the locus of the centre of a circle:

Ex. 150. Which has a given radius r and passes through a given point P.

Ex. 151. Which has a given radius r and touches a given line AB.

Ex. 152. Which passes through two given points P and Q.

Ex. 153. Which touches a given straight line AB at a given point P.

Ex. 154. Which touches each of two given parallels.

Ex. 155. Which touches each of two given intersecting lines.

Ex. 156. To find in a given line a point X which is equidistant from two given points.

The required point is the intersection of the given line with the perpendicular bisector of the line joining the two given points (§ 160).

Ex. 157. To find a point X equidistant from three given points.

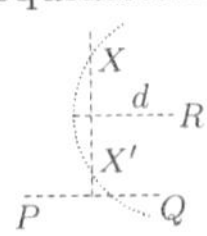

Ex. 158. To find a point X equidistant from two given points and at a given distance from a third given point.

Ex. 159. To construct a circle which has a given radius and passes through two given points.

Ex. 160. To find a point X at given distances from two given points.

Ex. 161. To construct a circle which has its centre in a given line and passes through two given points.

Ex. 162. To find a point X equidistant from two given points and also equidistant from two given intersecting lines (§§ 160 and 162).

Ex. 163. To find a point X equidistant from two given points and also equidistant from two given parallel lines.

Ex. 164. To find a point X equidistant from two given intersecting lines and also equidistant from two given parallels.

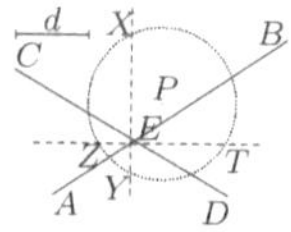

Ex. 165. To find a point X equidistant from two given intersecting lines and at a given distance from a given point.

Ex. 166. To find a point X which lies in one side of a given triangle and is equidistant from the other two sides.

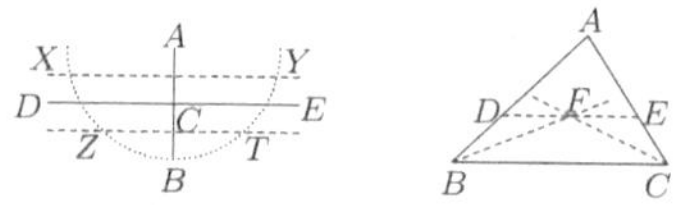

Ex. 167. A straight railway passes two miles from a town. A place is four miles from the town and one mile from the railway. To find by construction the places that answer this description.

Ex. 168. In a triangle ABC, to draw DE parallel to the base BC, cutting the sides of the triangle in D and E, so that DE shall equal $DB+EC$ (§ 162).

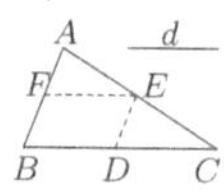

Ex. 169. To draw through two sides of a triangle a line parallel to the third side so that the part intercepted between the sides shall have a given length.

Take $BD = d$.

Ex. 170. Prove that the locus of the vertex of a right triangle, having a given hypotenuse as base, is the circumference described upon the given hypotenuse as diameter (§ 290).

Ex. 171. Prove that the locus of the vertex of a triangle, having a given base and a given angle at the vertex, is the arc which forms with the base a segment capable of containing the given angle (§ 318).

Ex. 172. Find the locus of the middle point of a chord of a given length that can be drawn in a given circle.

Ex. 173. Find the locus of the middle point of a chord drawn from a given point in a given circumference.

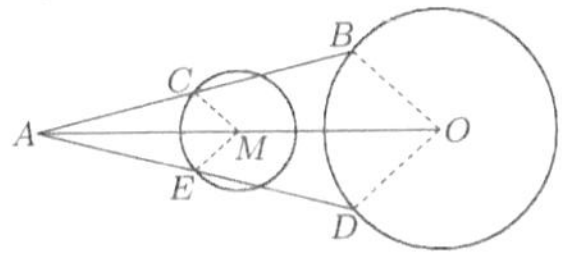

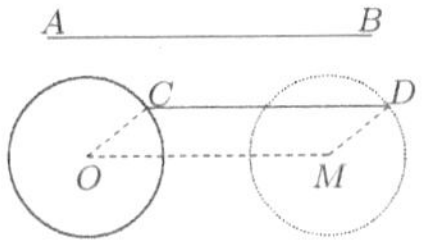

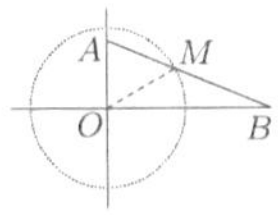

Ex. 174. Find the locus of the middle point of a straight line drawn from a given exterior point to a given circumference.

Ex. 175. A straight line moves so that it remains parallel to a given line, and touches at one end a given circumference. Find the locus of the other end.

Ex. 176. A straight rod moves so that its ends constantly touch two fixed rods which are perpendicular to each other. Find the locus of its middle point.

Ex. 177. In a given circle let AOB be a diameter, OC any radius, CD the perpendicular from C to AB. Upon OC take OM equal to CD. Find the locus of the point M as OC turns about O.

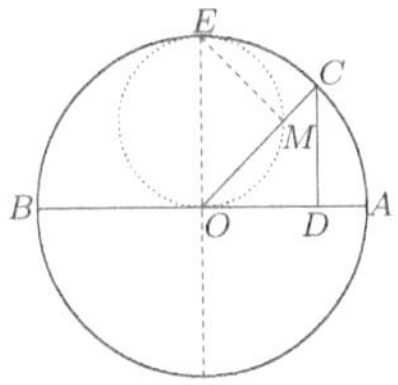

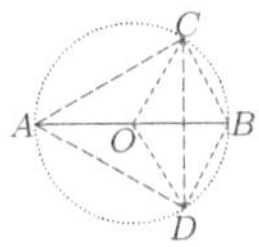

Ex. 178. To construct an equilateral triangle, having given the radius of the circumscribed circle.

To construct on isosceles triangle, having given:

Ex. 179. The angle at the vertex and the base (§ 160 and § 318).

Ex. 180. The base and the radius of the circumscribed circle.

Ex. 181. The base and the radius of the inscribed circle.

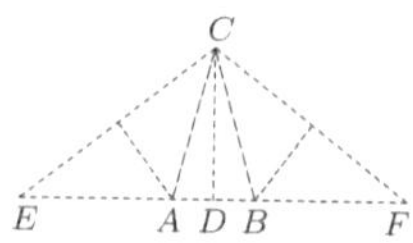

Ex. 182. The perimeter and the altitude.

Let ABC be the required, EF the given perimeter. The altitude CD passes through the middle of EF, and the $_sAEC$, BFC are isosceles.

To construct a right triangle, having given:

Ex. 183. The hypotenuse and one leg.

Ex. 184. One leg and the altitude upon the hypotenuse.

Ex. 185. The median and the altitude drawn from the vertex of the right angle.

Ex. 186. The hypotenuse and the altitude upon the hypotenuse.

Ex. 187. The radius of the inscribed circle and one leg.

Ex. 188. The radius of the inscribed circle and an acute angle.

Ex. 189. An acute angle and the sum of the legs.

Ex. 190. An acute angle and the difference of the legs.

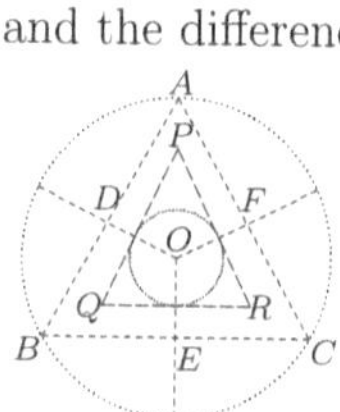

Ex. 191. To construct an equilateral triangle, having given the radius of the inscribed circle.

To construct a triangle, having given:

Ex. 192. The base, the altitude, and an angle at the base.

Ex. 193. The base, the altitude, and the at the vertex.

Ex. 194. The base, the corresponding median, and the at the vertex.

Ex. 195. The perimeter and the angles.

Ex. 196. One side, an adjacent , and the sum of the other sides.

To construct a triangle, having given:

Ex. 197. One side, an adjacent , and the difference of the other sides.

Ex. 198. The sum of two sides and the angles.

Ex. 199. One side, an adjacent , and the radius of the circumscribed circle.

Ex. 200. The angles and the radius of the circumscribed circle.

Ex. 201. The angles and the radius of the inscribed circle.

Ex. 202. An angle, and the bisector and the altitude drawn from the vertex of the given angle.

Ex. 203. Two sides and the median corresponding to the other side.

Ex. 204. The three medians.

To construct a square, having given:

Ex. 205. The diagonal.

Ex. 206. The sum of the diagonal and one side.

Let $ABCD$ be the square required, CA the diagonal. Produce CA, making $AE = AB$. $_{s}ABC$ and ABE are isosceles and $BAC =$ $BCA = 45°$.

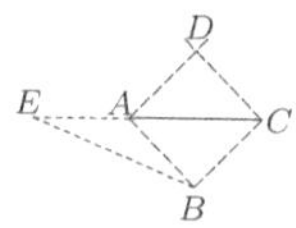

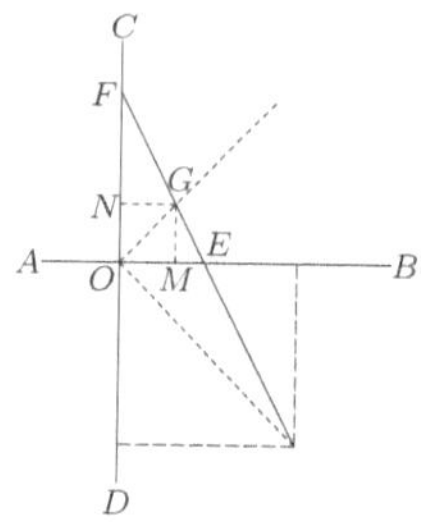

Ex. 207. Given two perpendiculars, AB and CD, intersecting in O, and a straight line intersecting these perpendiculars in E and F; to construct a square, one of whose angles shall coincide with one of the right angles at O, and the vertex of the opposite angle of the square shall lie in EF. (Two solutions.)

To construct a rectangle, having given:

Ex. 208. One side and the angle between the diagonals.

Ex. 209. The perimeter and the diagonal.

Ex. 210. The perimeter and the angle between the diagonals.

Ex. 211. The difference of two adjacent sides and the angle between the diagonals.

To construct a rhombus, having given:

Ex. 212. The two diagonals.

Ex. 213. One side and the radius of the inscribed circle.

Ex. 214. One angle and the radius of the inscribed circle.

Ex. 215. One angle and one of the diagonals.

To construct a rhomboid, having given:

Ex. 216. One side and the two diagonals.

Ex. 217. The diagonals and the angle between them.

Ex. 218. One side, one angle, and one diagonal.

Ex. 219. The base, the altitude, and one angle.

To construct an isosceles trapezoid, having given:

Ex. 220. The bases and one angle.

Ex. 221. The bases and the altitude.

Ex. 222. The bases and the diagonal.

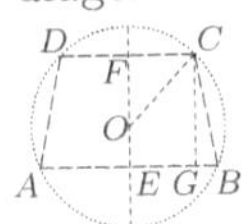

Ex. 223. The bases and the radius of the circumscribed circle.

Let $ABCD$ be the isosceles trapezoid required, O the centre of the circumscribed $\odot$. A diameter to AB is to CD, and bisects both AB and CD. Draw $CG \parallel$ to FE. Then $EG = FC = \frac{1}{2}DC$.

To construct a trapezoid, having given:

Ex. 224. The four sides.

Ex. 225. The two bases and the two diagonals.

Ex. 226. The bases, one diagonal, and the between the diagonals.

To construct a circle which has the radius r and which also:

Ex. 227. Touches each of two intersecting lines AB and CD.

Ex. 228. Touches a given line AB and a given circle K.

Ex. 229. Passes through a given point P and touches a given line AB.

Ex. 230. Passes through a given point P and touches a given circle K.

To construct a circle which shall:

Ex. 231. Touch two given parallels and pass through a given point P.

Ex. 232. Touch three given lines two of which are parallel.

Ex. 233. Touch a given line AB at P and pass through a given point Q.

Ex. 234. Touch a given circle at P and pass through a given point Q.

Ex. 235. Touch two given lines and touch one of them at a given point P.

Ex. 236. Touch a given line and touch a given circle at a point P.

Ex. 237. Touch a given line AB at P and also touch a given circle.

Ex. 238. To inscribe a circle in a given sector.

Ex. 239. To construct within a given circle three equal circles, so that each shall touch the other two and also the given circle.

Ex. 240. To describe circles about the vertices of a given triangle as centres, so that each shall touch the two others.

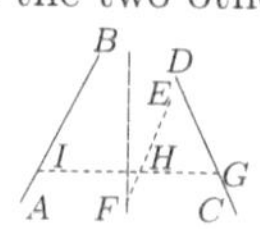

Ex. 241. To bisect the angle formed by two lines, without producing the lines to their point of intersection.

Draw any line EF $\parallel$ to BA. Take $EG = EH$, and produce GH to meet BA at I. Draw the bisector of GI.

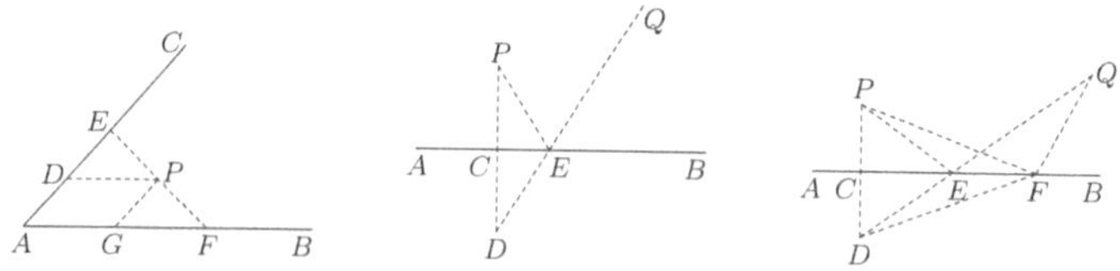

Ex. 242. To draw through a given point P between the sides of an angle BAC a line terminated by the sides of the angle and bisected at P.

Ex. 243. Given two points P, Q, and a line AB; to draw lines from P and Q which shall meet on AB and make equal angles with AB.

Make use of the point which forms with P a pair of points symmetrical with respect to AB.

Ex. 244. To find the shortest path from P to Q which shall touch a line AB.

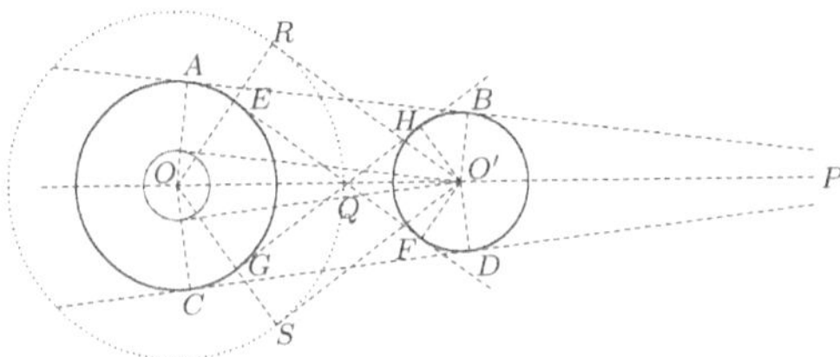

Ex. 245. To draw a common tangent to two given circles.

Let r and r' denote the radii of the circles, O and O' their centres. With centre O and radius $r - r'$ describe a $\odot$. From O' draw the tangents $O'M$, $O'N$. Produce OM and ON to meet the circumference at A and C. Draw the radii $O'B$ and $O'D$ $\|$, respectively, to OA and OC. Draw AB and CD.

To draw the internal tangents use an auxiliary $\odot$ of radius $r + r'$.

BOOK III. PROPORTION. SIMILAR POLYGONS.

THE THEORY OF PROPORTION.

323. A **proportion** is an expression of equality between two equal ratios; and is written in one of the following forms:

$$a : b = c : d; \qquad a : b :: c : d; \qquad \frac{a}{b} = \frac{c}{d}.$$

This proportion is read, "a is to b as c is to d"; or "the ratio of a to b is equal to the ratio of c to d."

324. The **terms** of a proportion are the four quantities compared; the *first* and *third* terms are called the **antecedents**, the *second* and *fourth* terms, the **consequents**; the *first* and *fourth* terms, the **extremes**, the *second* and *third* terms, the **means**.

Thus, in the proportion $a : b = c : d$; a and c are the antecedents, b and d the consequents, a and d the extremes, b and c the means.

325. The fourth proportional to three given quantities is the fourth term of the proportion which has for its first three terms the three given quantities *taken in order.*

Thus, d is the fourth proportional to a, b, and c in the proportion

$$a : b = c : d.$$

326. The quantities a, b, c, d, e, are said to be in **continued proportion**, if $a : b = b : c = c : d = d : e$.

If three quantities are in continued proportion, the second is called the **mean proportional** between the other two, and the third is called the **third proportional** to the other two.

Thus, in the proportion $a : b = b : c$; b is the mean proportional between a and c; and c is the third proportional to a and b.

PROPOSITION I. THEOREM.

327. *In every proportion the product of the extremes is equal to the product of the means.*

Let

$$\mathbf{a : b = c : d}.$$

Then

$$\frac{a}{b} = \frac{c}{d}. \qquad \S\ 323$$

Whence

$$ad = bc. \qquad \text{Q.E.D.}$$

PROPOSITION II. THEOREM.

328. *The mean proportional between two quantities is equal to the square root of their product.*

Let

$$\mathbf{a : b = b : c}.$$

Then

$$b^2 = ac. \qquad \S\ 327$$

Whence, extracting the square root,

$$b = \sqrt{ac}. \qquad \text{Q.E.D.}$$

PROPOSITION III. THEOREM.

329. *If the product of two quantities is equal to the product of two others, either two may be made the extremes of the proportion in which the other two are made the means.*

Let

$$\mathbf{ad = bc}.$$

To prove that

$$a : b = c : d$$

Divide both members of the given equation by bd.

Then

$$\frac{a}{b} = \frac{c}{d}.$$

Or

$$a : b = c : d. \qquad \text{Q.E.D.}$$

Proposition IV. Theorem.

330. *If four quantities are in proportion, they are in proportion by* **alternation***; that is, the first term is to the third as the second is to the fourth.*
Let

$$\mathbf{a} : \mathbf{b} = \mathbf{c} : \mathbf{d}.$$

To prove that

$$a : c = b : d.$$

Now

$$ad = bc. \qquad \S\ 327$$

Divide each member of the equation by cd.
Then

$$\frac{a}{c} = \frac{b}{d}.$$

Or

$$a : c = b : d. \qquad \text{Q.E.D.}$$

Proposition V. Theorem.

331. *If four quantities are in proportion, they are in proportion by* **inversion***; that is, the second term is to the first as the fourth is to the third.*
Let

$$\mathbf{a} : \mathbf{b} = \mathbf{c} : \mathbf{d}.$$

To prove that

$$b : a = d : c.$$

Now

$$bc = ad. \qquad \S\ 327$$

Divide each member of the equation by ac.
Then

$$\frac{b}{a} = \frac{d}{c}.$$

Or

$$b : a = d : c. \qquad \text{Q.E.D.}$$

PROPOSITION VI. THEOREM.

332. *If four quantities are in proportion, they are in proportion by* **composition** *that is, the sum of the first two terms is to the second term as the sum of the last two terms is to the fourth term.*

Let

$$\mathbf{a} : \mathbf{b} = \mathbf{c} : \mathbf{d}.$$

To prove that

$$a + b : b = c + d : d.$$

Now

$$\frac{a}{b} = \frac{c}{d}.$$

Then

$$\frac{a}{b} + 1 = \frac{c}{d} + 1;$$

that is,

$$\frac{a+b}{b} = \frac{c+d}{d}.$$

Or

$$a + b : b = c + d : d.$$

In like manner

$$a + b : a = c + d : c.$$

Q.E.D.

PROPOSITION VII. THEOREM.

333. *If four quantities are in proportion, they are in proportion by* **division**; *that is, the difference of the first two terms is to the second term as the difference of the last two terms is to the fourth term.*

Let

$$\mathbf{a} : \mathbf{b} = \mathbf{c} : \mathbf{d}.$$

To prove that

$$a - b : b = c - d : d.$$

Now

$$\frac{a}{b} = \frac{c}{d}.$$

Then

$$\frac{a}{b} - 1 = \frac{c}{d} - 1;$$

that is,

$$\frac{a-b}{b} = \frac{c-d}{d}.$$

Or

$$a - b : b = c - d : d.$$

In like manner

$$a - b : a = c - d : c.$$

Q.E.D.

PROPOSITION VIII. THEOREM.

334. *If four quantities are in proportion, they are in proportion by* **composition and division***; that is, the sum of the first two terms is to their difference as the sum of the last two terms is to their difference.*

Let

$$\mathbf{a : b = c : d}$$

Then

$$\frac{a+b}{a} = \frac{c+d}{c}. \qquad \S\ 332$$

And

$$\frac{a-b}{a} = \frac{c-d}{c}. \qquad \S\ 333$$

Divide,

$$\frac{a+b}{a-b} = \frac{c+d}{c-d}.$$

Or

$$a+b : a-b = c+d : c-d. \qquad \text{Q.E.D.}$$

PROPOSITION IX. THEOREM.

335. *In a series of equal ratios, the sum of the antecedents is to the sum of the consequents as any antecedent is to its consequent.*

Let

$$\mathbf{a} : \mathbf{b} = \mathbf{c} : \mathbf{d} = \mathbf{e} : \mathbf{f} = \mathbf{g} : \mathbf{h}.$$

To prove that $a + c + e + g : b + d + f + h = a : b$.

Let

$$r = \frac{a}{b} = \frac{c}{d} = \frac{e}{f} = \frac{g}{h}.$$

Then

$$a = br,\ c = dr,\ e = fr,\ g = hr.$$

And

$$a + c + e + g = (b + d + f + h)r.$$

Divide by $(b + d + f + h)$.

Then

$$\frac{a + c + e + g}{b + d + f + h} = r = \frac{a}{b}.$$

Or

$$a + c + e + g : b + d + f + h = a : b.$$

Q.E.D.

Proposition X. Theorem.

336. *The products of the corresponding terms of two or more proportions are in proportion.*

Let

$$\mathbf{a : b = c : d,\ e : f = g : h,\ k : l = m : n.}$$

To prove that

$$aek : bfl = cgm : dhn.$$

Now

$$\frac{a}{b} = \frac{c}{d}, \frac{e}{f} = \frac{g}{h}, \frac{k}{l} = \frac{m}{n}.$$

The products of the first members and of the second members of these equations give

$$\frac{aek}{bfl} = \frac{cgm}{dhn}.$$

Or

$$aek : bfl = cgm : dhn. \qquad \text{Q.E.D.}$$

337. Cor. *If three quantities are in continued proportion, the first is to the third as the square of the first is to the square of the second.*

Proposition XI. Theorem.

338. *Like powers of the terms of a proportion are in proportion.*

Let

$$\mathbf{a : b = c : d.}$$

To prove that

$$a^n : b^n = c^n : d^n.$$

Now

$$\frac{a}{b} = \frac{c}{d}.$$

Raise to the nth power,

$$\frac{a^n}{b^n} = \frac{c^n}{d^n}.$$

Or

$$a^n : b^n = c^n : d^n. \qquad \text{Q.E.D.}$$

339. DEF. **Equimultiples** of two quantities are the products obtained by multiplying each of them by the same number. Thus, ma and mb are equimultiples of a and b.

PROPOSITION XII. THEOREM.

340. *Equimultiples of two quantities are in the same ratio as the quantities themselves.*

Let a and b be any two quantities.

To prove that

$$ma : mb = a : b.$$

Now

$$\frac{a}{b} = \frac{a}{b}.$$

Multiply both terms of the first fraction by m.

Then

$$\frac{ma}{mb} = \frac{a}{b}.$$

Or

$$ma : mb = a : b.$$

Q.E.D.

341. SCHOLIUM. In the treatment of proportion, it is assumed that the *quantities* involved are expressed by their *numerical measures.* It is evident that the ratio of two quantities of the same kind may be represented by a fraction, if the two quantities are expressed in *integers* in terms of a *common unit.* If there is no unit in terms of which both quantities can be expressed in *integers*, it is still possible by taking the unit of measure sufficiently small to find a fraction that will represent the ratio *to any required degree of accuracy.* § 269

If we speak of the product of two quantities, it must be understood that we mean simply *the product of the numbers which represent them when they are expressed in terms of a common unit.*

In order that four quantities, a, b, c, d, may form a proportion, a and b must be quantities of the same kind; and c and d must be quantities of the same kind; though c and d need not be of the same kind as a and b. In alternation, however, *the four quantities must be of the same kind.*

Proposition XIII. Theorem.

342. *If a line is drawn through two sides of a triangle parallel to the third side, it divides those sides proportionally.*

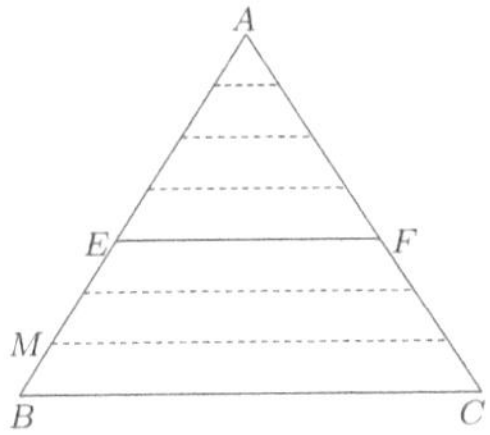

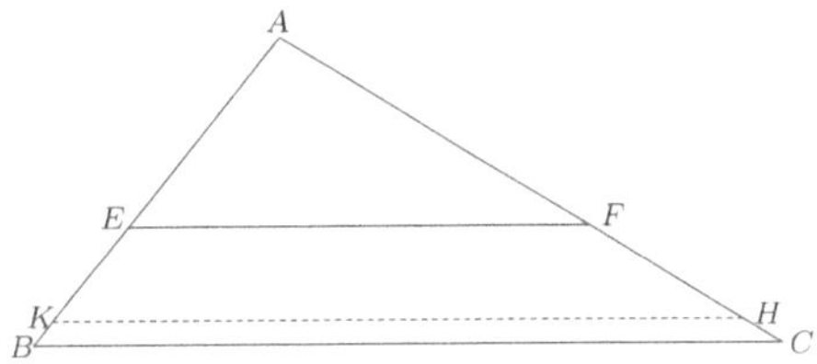

In the triangle ABC, let EF be drawn parallel to BC.

To prove that

$$EB : AE = FC : AF.$$

Case 1. *When AE and EB* (Fig. 1) *are commensurable.*

Proof. Find a common measure of AE and EB, as MB.

Let MB be contained m times in EB, and n times in AE.

Then

$$EB : AE = m : n.$$

At the points of division on BE and AE draw lines $\parallel$ to BC. These lines will divide AC into $m+n$ equal parts, of which FC will contain m, and AF will contain n. § 187

$$FC : AF = m : n.$$

$$EB : AE = FC : AF. \qquad \text{Ax. 1}$$

Case 2. *When AE and EB* (Fig. 2) *are incommensurable.*

Proof. Divide AE into any number of equal parts, and apply one of these parts to EB as many times as EB will contain it.

Since AE and EB are incommensurable, a certain number of these parts will extend from E to some point K, leaving a remainder KB less than one of these parts. Draw $KH \parallel BC$.

Then

$$EK : AE = FH : AF \qquad \text{Case 1}$$

By increasing the *number* of equal parts into which AE is divided, we can make the *length* of each part less than any assigned value, however small, but not zero.

Hence, KB, which is less than one of these equal parts, has zero for a limit. § 275

And the corresponding segment HC has zero for a limit.

Therefore, EK approaches EB as a limit, § 271
and FH approaches FC as a limit.

$$\text{the variable } \frac{EK}{AE} \text{ approaches } \frac{EB}{AE} \text{ as a limit,}$$ § 280

$$\text{and the variable } \frac{FH}{AF} \text{ approaches } \frac{FC}{AF} \text{ as a limit.}$$

But

$$\frac{EK}{AE} \text{ is constantly equal to } \frac{FH}{AF}$$ Case 1

$$\frac{EB}{AE} = \frac{FC}{AF}.$$ § 284

Q.E.D.

343. Cor. 1. *One side of a triangle is to either part cut off by a straight line parallel to the base as the other side is to the corresponding part.*

For

$$AE : EB = AF : FC.$$

By composition,

$$AE + EB : AE = AF + FC : AF.$$ § 332

Or

$$AB : AE = AC : AF.$$

344. Cor. 2. *If two lines are cut by any number of parallels the corresponding intercepts are proportional.*

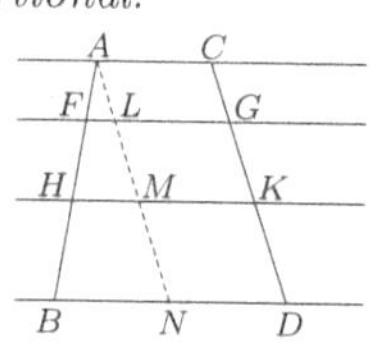

Draw $AN \parallel$ to CD. Then

$$AL = CG,\ LM = GK,\ MN = KD.$$ § 180

Now

$$AH : AM = AF : AL = FH : LM$$
$$= HB : MN.$$ § 343

Or

$$AF : CG = FH : GK = HB : KD.$$

PROPOSITION XIV. THEOREM.

345. *If a straight line divides two sides of a triangle proportionally, it is parallel to the third side.*

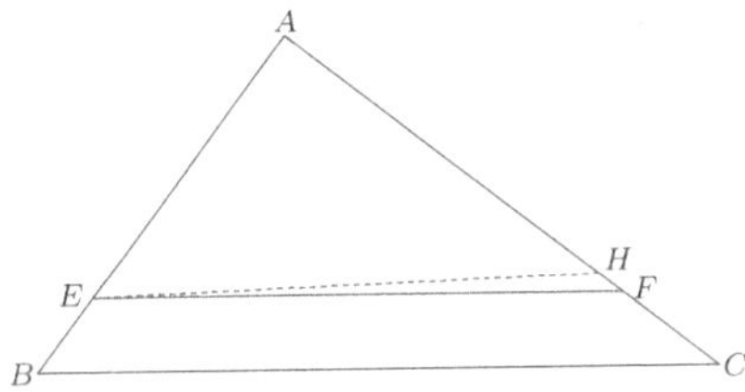

In the triangle ABC, let EF be drawn so that

$$\frac{AB}{AE} = \frac{AC}{AF}.$$

To prove that

$$EF \text{ is } \parallel \text{ to } BC.$$

Proof.

$$\text{From } E \text{ draw } EH \parallel \text{ to } BC.$$

Then

$$AB : AE = AC : AH, \qquad \S\ 343$$

(*one side of a triangle is to either part cut off by a line parallel to the base as the other side to the corresponding part*).

But

$$AB : AE = AC : AF. \qquad \text{Hyp.}$$

$$\therefore AC : AF = AC : AH. \qquad \text{Ax. 1}$$

$$\therefore AF = AH.$$

$$\therefore EF \text{ and } EH \text{ coincide.} \qquad \S\ 47$$

But

$$EH \text{ is } \parallel \text{ to } BC. \qquad \text{Const.}$$

$$\therefore EF, \text{ which coincides with } EH, \text{ is } \parallel \text{ to } BC. \qquad \text{Q.E.D.}$$

Ex. 246. Find the fourth proportional to 91, 65, and 133.

Ex. 247. Find the mean proportional between 39 and 351.

Ex. 248. Find the third proportional to 54 and 3.

346. If a given line AB is divided at M, a point between the extremities A and B, it is said to be divided **internally** into the segments MA and MB; and if it is divided at M', a point in the prolongation of AB, it is said to be divided **externally** into the segments $M'A$ and $M'B$.

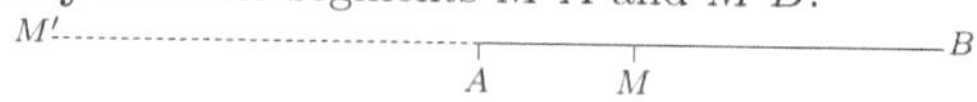

In either case the segments are the *distances* from the point of division to the extremities of the line. If the line is divided internally, the *sum* of the segments is equal to the line; and if the line is divided externally, the *difference* of the segments is equal to the line.

Suppose it is required to divide the given line AB **internally and externally in the same ratio**; as, for example, the ratio of the two numbers 3 and 5.

We divide AB into $5 + 3$, or 8, equal parts, and take 3 parts from A; we then have the point M, such that

$$MA : MB = 3 : 5. \qquad (1)$$

Secondly, we divide AB into $5 - 3$, or 2, equal parts, and lay off on the prolongation of AB, to the left of A, three of these equal parts; we then have the point M', such that

$$M'A : M'B = 3 : 5. \qquad (2)$$

Comparing (1) and (2),

$$MA : MB = M'A : M'B.$$

347. DEF. If a given straight line is divided internally and externally into segments having the same ratio, the line is said to be **divided harmonically**.

Proposition XV. Theorem.

348. *The bisector of an angle of a triangle divides the opposite side into segments which are proportional to the adjacent sides.*

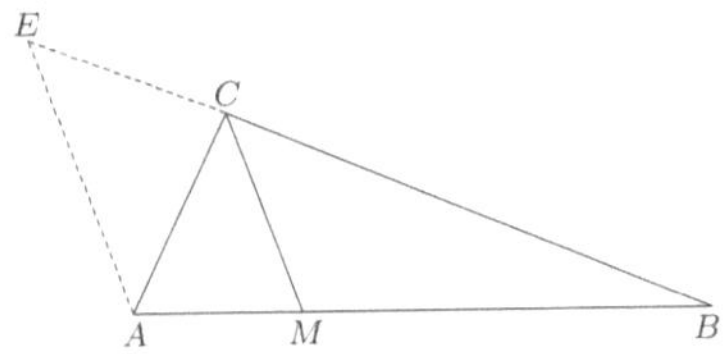

Let CM bisect the angle C of the triangle CAB.

To prove that

$$MA : MB = CA : CB.$$

Proof. Draw $AE \parallel$ to MC, meeting BC produced at E.

Then

$$MA : MB = CE : CB,$$ § 342

(*if a line is drawn through two sides of a   parallel to the third side, it divides those sides proportionally*).

Also,

$$ACM = \quad CAE,$$ § 110

(*being alt.-int.* *s of $\parallel$ lines*);

and

$$BCM = \quad CEA,$$ § 112

(*being ext.-int.* *s of $\parallel$ lines*).

But

$$ACM = \quad BCM.$$ Hyp.

$$CAE = \quad CEA.$$ Ax. 1

$$CE = CA.$$ § 147

Put CA for its equal, CE, in the first proportion.

Then

$$MA : MB = CA : CB.$$ Q.E.D.

Ex. 249. In a triangle ABC, $AB = 12$, $AC = 14$, $BC = 13$. Find the segments of BC made by the bisector of the angle A.

Ex. 250. In a triangle CAB, $CA = 6$, $CB = 12$, $AB = 15$. Find the segments of AB made by the bisector of the angle C.

PROPOSITION XVI. THEOREM.

349. *The bisector of an exterior angle of a triangle divides the opposite side externally into segments which are proportional to the adjacent sides.*

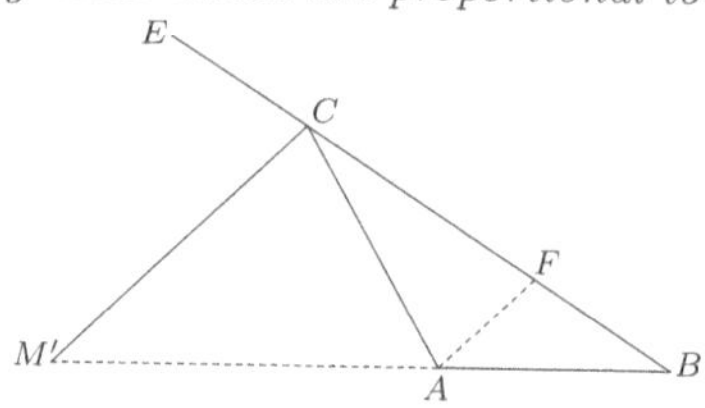

Let CM' bisect the exterior angle ACE of the triangle CAB, and meet BA produced at M'.

To prove that

$$M'A : M'B = CA : CB.$$

Proof.

Draw $AF \parallel$ to $M'C$, meeting BC at F.

Then

$$M'A : M'B \; CF : CB. \qquad \S\ 343$$

Now

$$M'CE = \quad AFC, \qquad \S\ 112$$

and

$$M'CA = \quad CAF, \qquad \S\ 110$$

(*being alt.-int.* $_s$ *of* $\parallel$ *lines*).

But

$$M'CE = \quad M'CA. \qquad \text{Hyp.}$$

$$AFC = \quad CAF. \qquad \text{Ax. 1}$$

$$CA = CF. \qquad \S\ 147$$

Put CA for its equal, CF, in the first proportion.

Then

$$M'A : M'B = CA : CB. \qquad \text{Q.E.D.}$$

Question. To what case does this theorem not apply? (See Ex. 41, page 79.)

350. Cor. *The bisectors of an interior angle and an exterior angle at one vertex of a triangle meeting the opposite side divide that side harmonically.* § 347

SIMILAR POLYGONS.

351. DEF. **Similar polygons** are polygons that have their homologous angles equal, and their homologous sides proportional.

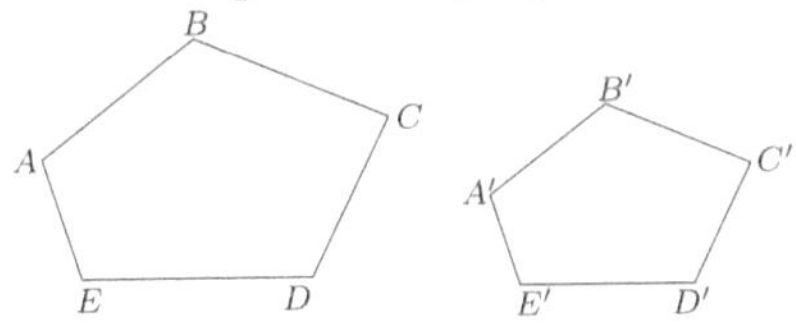

Thus, the polygons $ABCDE$ and $A'B'C'D'E'$ are similar, if the ${}_sA$, B, C, etc., are equal, respectively, to the ${}_sA'$, B', C', etc., and if

$$AB : A'B' = BC : B'C' = CD : C'D', \text{ etc.}$$

352. DEF. **Homologous lines** in similar polygons are lines similarly situated.

353. DEF. The ratio of any two homologous lines in similar polygons, is called the **ratio of similitude** of the polygons.

The primary idea of similarity is **likeness of form**. The two conditions *necessary* to similarity are:

1. *For every angle in one of the figures there must be an equal angle in the other.*
2. *The homologous sides must be proportional.*

Thus, Q and Q' are not similar; the homologous sides are proportional, but the homologous angles are not equal. Also R and R' are not similar; the homologous angles are equal, but the sides are not proportional.

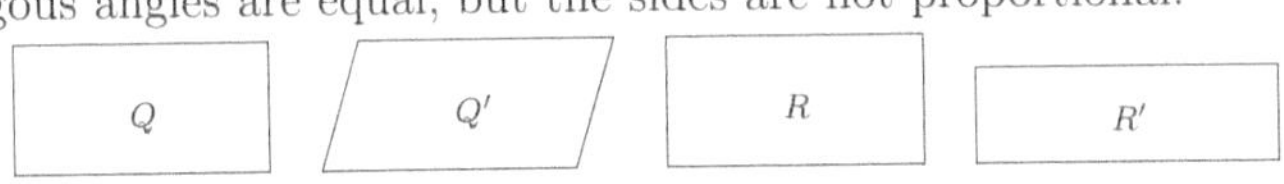

In the case of *triangles*, either condition involves the other (see § 354 and § 358).

Proposition XVII. Theorem.

354. *Two mutually equiangular triangles are similar.*

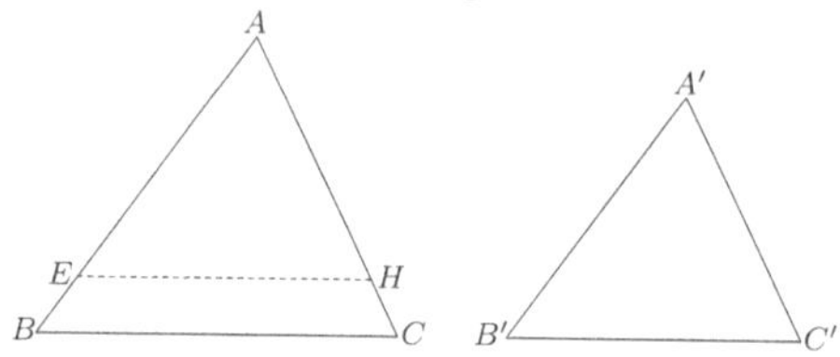

In the triangles ABC and $A'B'C'$, let the angles A, B, C be equal to the angles A', B', C', respectively.

To prove that the $_s ABC$ and $A'B'C'$ are similar.

Since the $_s$ are mutually equiangular, we have only to prove that

$$AB : A'B' = AC : A'C' = BC : B'C'. \qquad \S\ 351$$

Proof. Place the $A'B'C'$ on the ABC so that A' shall coincide with its equal, the A; and $B'C'$ take the position EH.

Then

$$AEH = B \qquad \text{Hyp.}$$

$$EH \text{ is } \parallel \text{ to } BC. \qquad \S\ 114$$

$$AB : AE = AC : AH. \qquad \S\ 343$$

That is,

$$AB : A'B' = AC : A'C'.$$

Similarly, by placing $A'B'C'$ on ABC, so that B' shall coincide with its equal, the B, we may prove that

$$AB : A'B' = BC : B'C' \qquad \text{Q.E.D.}$$

355. Cor. 1. *Two triangles are similar if two angles of the one are equal, respectively, to two angles of the other.*

356. Cor. 2. *Two right triangles are similar if an acute angle of the one is equal to an acute angle of the other.*

PROPOSITION XVIII. THEOREM.

357. *If two triangles have an angle of the one equal to an angle of the other, and the including sides proportional, they are similar.*

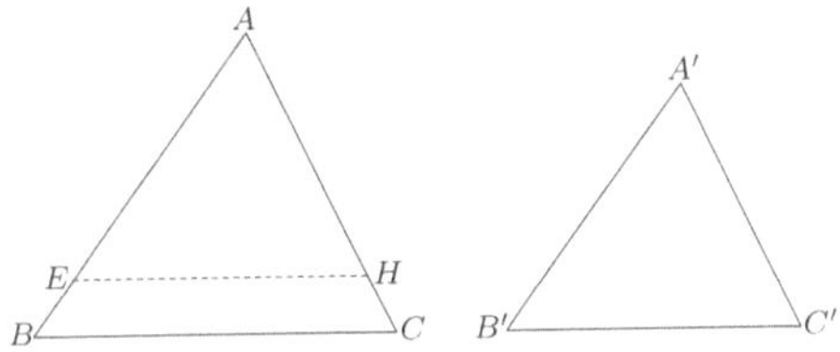

In the triangles ABC and $A'B'C'$, let $A =$ A', and let

$$\mathbf{AB : A'B' = AC : A'C'.}$$

To prove that the $_s ABC$ *and* $A'B'C'$ *are similar.*

In this case we prove the $_s$ similar by proving them mutually equiangular.

Proof. Place the $A'B'C'$ on the ABC, so that the A' shall coincide with its equal, the A.

Then the $A'B'C'$ will take the position of AEH.

Now
$$\frac{AB}{A'B'} = \frac{AC}{A'C'}. \qquad \text{Hyp.}$$

That is,
$$\frac{AB}{AE} = \frac{AC}{AH}.$$

EH is $\|$ to BC, § 345

(*if a line divides two sides of a proportionally, it is* $\|$ *to the third side*).

$AEH =$ B, and $AHE =$ C. § 112

AEH is similar to ABC. § 354

$A'B'C'$ is similar to ABC. Q.E.D.

PROPOSITION XIX. THEOREM.

358. *If two triangles have their sides respectively proportional, they are similar.*

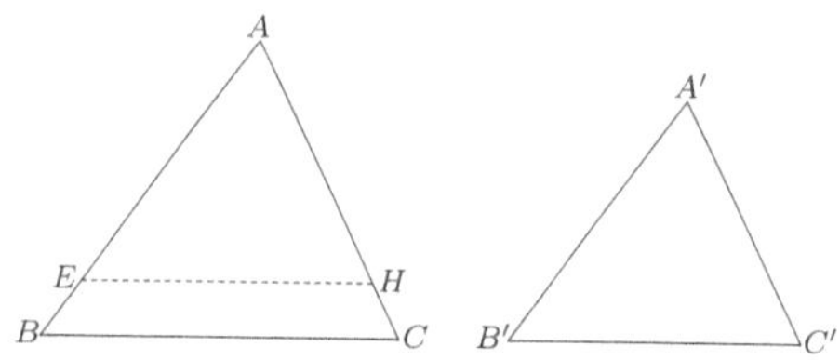

In the triangles ABC and $A'B'C'$, let

$$\mathbf{AB : A'B' = AC : A'C' = BC : B'C'}.$$

To prove that the $_sABC$ *and* $A'B'C'$ *are similar.*

Proof. Upon AB take AE equal to $A'B'$, and upon AC take AH equal to $A'C'$; and draw EH.

Now

$$AB : A'B' = AC : A'C'. \quad \text{Hyp.}$$

Or, since

$$AE = A'B' \text{ and } AH = A'C',$$

$$AB : AE = AC : AH.$$

$$_sABC \text{ and } AEH \text{ are similar.} \quad \S\ 357$$

$$AB : AE = BC : EH; \quad \S\ 351$$

that is,

$$AB : A'B' = BC : EH.$$

But

$$AB : A'B' = BC : B'C'. \quad \text{Hyp.}$$

$$BC : EH = BC : B'C'. \quad \text{Ax. 1}$$

$$EH = B'C'.$$

Hence, the $_sAEH$ and $A'B'C'$ are equal. § 150

But

$$AEH \text{ is similar to } ABC.$$

$$A'B'C' \text{ is similar to } ABC. \quad \text{Q.E.D.}$$

PROPOSITION XX. THEOREM.

359. *Two triangles which have their sides respectively parallel, or respectively perpendicular, are similar.*

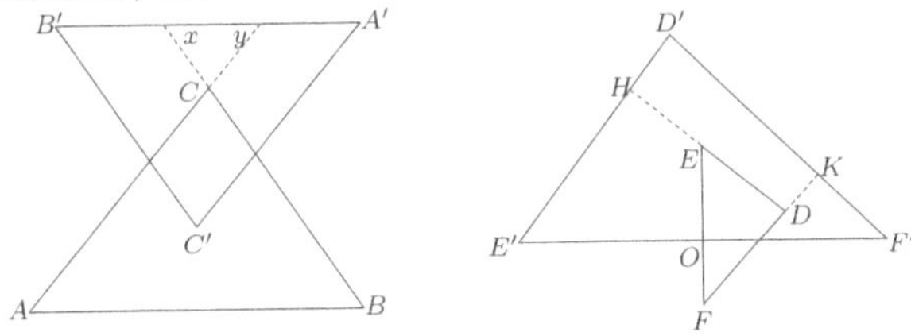

Let ABC and $A'B'C'$ have their sides respectively parallel; and DEF and $D'E'F'$ have their sides respectively perpendicular.

To prove that the $_s ABC$ and $A'B'C'$ are similar; and that the $_s DEF$ and $D'E'F'$ are similar.

Proof. 1. Prolong BC and AC to $B'A'$, forming $_s x$ and y.

Then $B' = x$ (§ 112), and $B = x$. § 110

Therefore,

$B' = B$ Ax. 1

In like manner,

$A' = A$.

Therefore, $A'B'C'$ is similar to ABC. § 355

2. Prolong DE and FD to meet $D'E'$ at H and $D'F'$ at K.

The quadrilateral $EHE'O$ has $_s EHE'$ and $E'OE$ right angles, by hypothesis. Therefore,

E' and OEH are supplementary. § 206

But

DEF and OEH are supplementary. § 86

Therefore, $DEF = E'$. § 85

In like manner,

$EDF = D'$.

Therefore, DEF is similar to $D'E'F'$. § 355

Q.E.D.

360. COR. *The parallel sides and the perpendicular sides are homologous sides of the triangles.*

PROPOSITION XXI. THEOREM.

361. *The homologous altitudes of two similar triangles have the same ratio as any two homologous sides.*

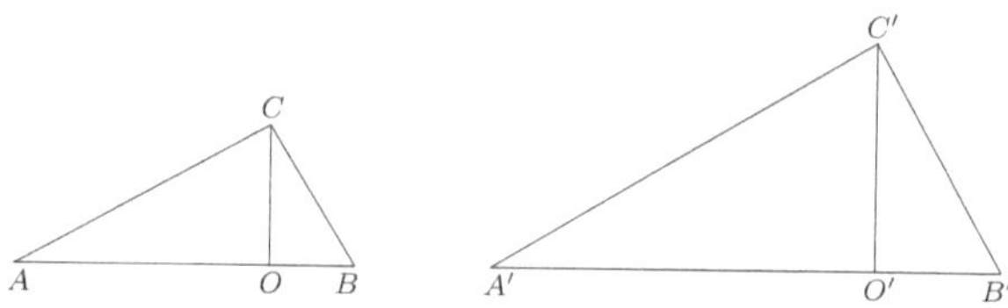

In the two similar triangles ABC and $A'B'C'$, let CO and $C'O'$ be homologous altitudes.

To prove that $\frac{CO}{C'O'} = \frac{AC}{A'C'} = \frac{AB}{A'B'} = \frac{BC}{B'C'}$.

Proof. In the rt. $_sCOA$ and $C'O'A'$,

$$A = A', \qquad \S\ 351$$

(*being homologous* $_s$ *of the similar* $_sABC$ *and* $A'B'C'$).

$_sCOA$ and $C'O'A'$ are similar, § 356

(*two rt.* $_s$ *having an acute* *of the one equal to an acute* *of the other are similar*).

$$\frac{CO}{C'O'} = \frac{AC}{A'C'}. \qquad \S\ 351$$

In the similar $_sABC$ and $A'B'C'$,

$$\frac{AC}{A'C'} = \frac{AB}{A'B'} = \frac{BC}{B'C'}. \qquad \S\ 351$$

Therefore,

$$\frac{CO}{C'O'} = \frac{AC}{A'C'} = \frac{AB}{A'B'} = \frac{BC}{B'C'}. \qquad \text{Q.E.D.}$$

Ex. 251. The base and altitude of a triangle are 7 feet 6 inches and 5 feet 6 inches, respectively. If the homologous base of a similar triangle is 5 feet 6 inches, find its homologous altitude.

PROPOSITION XXII. THEOREM.

362. *If two parallels are cut by three or more transversals that pass through the same point, the corresponding segments are proportional.*

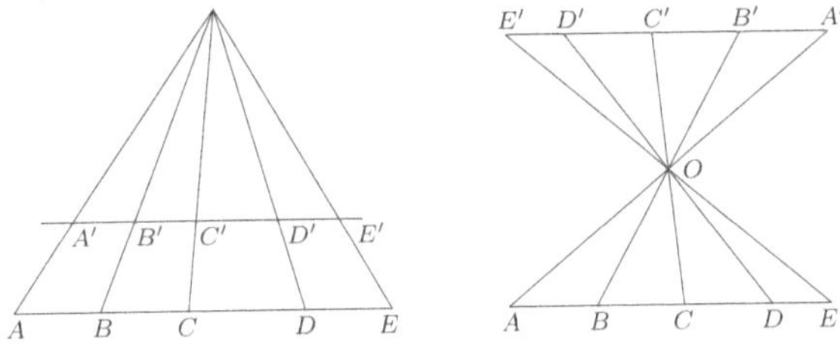

Let the two parallels AE and $A'E'$ be cut by the transversals OA, OB, OC, OD, OE in A, A', B, B', etc.

To prove that $\frac{AB}{A'B'} = \frac{BC}{B'C'} = \frac{CD}{C'D'} = \frac{DE}{D'E'}$.

Proof. Since $A'E'$ is $\parallel$ to AE, the pairs of $_{s}OAB$ and $OA'B'$, OBC and $OB'C'$, etc., are similar. § 354

$$\frac{AB}{A'B'} = \frac{OB}{OB'} \text{ and } \frac{BC}{B'C'} = \frac{OB}{OB'}. \qquad \S\ 351$$

(*homologous sides of similar $_s$ are proportional*).

$$\frac{AB}{A'B'} = \frac{BC}{B'C'}. \qquad \text{Ax. 1}$$

In a similar way it may be shown that

$$\frac{BC}{B'C'} = \frac{CD}{C'D'} \text{ and } \frac{CD}{C'D'} = \frac{DE}{D'E'}. \qquad \text{Q.E.D.}$$

NOTE. A condensed form of writing the above is

$$\frac{AB}{A'B'} = \left(\frac{OB}{OB'}\right) = \frac{BC}{B'C'} = \left(\frac{OC}{OC'}\right) = \frac{CD}{C'D'} = \left(\frac{OD}{OD'}\right) = \frac{DE}{D'E'}.$$

A parenthesis about a ratio signifies that this ratio is used to prove the equality of the ratios immediately preceding and following it.

Proposition XXIII. Theorem.

363. Conversely: *If three or more non-parallel straight lines intercept proportional segments upon two parallels, they pass through a common point.*

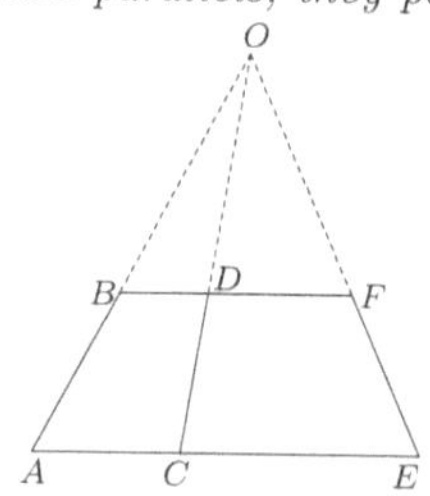

Let AB, CD, EF cut the parallels AE and BF so that

$$\mathbf{AC : BD = CE : DF.}$$

To prove that AB, CD, EF prolonged meet in a point.

Proof. Prolong AB and CD until they meet in O.

Draw OE.

Designate by F' the point where OE cuts BF.

Then

$$AC : BD = CE : DF'. \quad \S\ 362$$

But

$$AC : BD = CE : DF. \quad \text{Hyp.}$$

$$CE : DF' = CE : DF. \quad \text{Ax. 1}$$

$$DF' = DF.$$

F' coincides with F.

EF coincides with EF'. § 47

EF prolonged passes through O.

AB, CD, and EF prolonged meet in the point O. Q.E.D.

Proposition XXIV. Theorem.

364. *The perimeters of two similar polygons have the same ratio as any two homologous sides.*

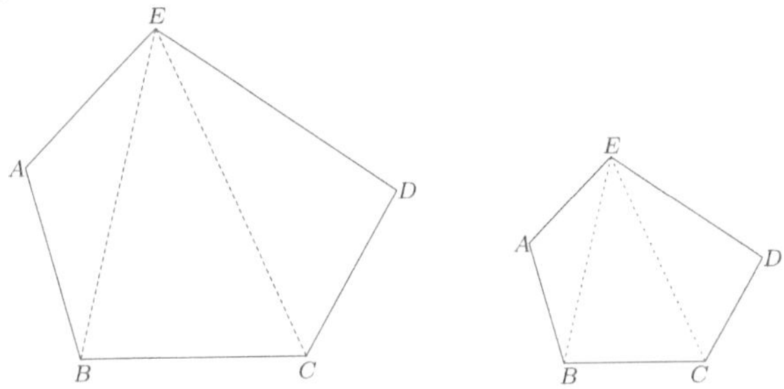

Let the two similar polygons be $ABCDE$ and $A'B'C'D'E'$, and let P and P' represent their perimeters.

To prove that

$$P : P' = AB : A'B'.$$

Proof.

$$AB : A'B' = BC : B'C' = CD : C'D', \text{ etc.} \qquad \S\ 351$$

$$AB + BC + \text{etc.} : A'B' + B'C' + \text{etc.} = AB : A'B', \qquad \S\ 335$$

(*in a series of equal ratios the sum of the antecedents is to the sum of the consequents as any antecedent is to its consequent*).

That is,

$$P : P' = AB : A'B'. \qquad \text{Q.E.D.}$$

Ex. 252. If the line joining the middle points of the bases of a trapezoid is produced, and the two legs are also produced, the three lines will meet in the same point.

Ex. 253. AB and AC are chords drawn from any point A in the circumference of a circle, and AD is a diameter. The tangent to the circle at D intersects AB and AC at E and F, respectively. Show that the triangles ABC and AEF are similar.

Ex. 254. AD and BE are two altitudes of the triangle CAB. Show that the triangles CED and CAB are similar.

Ex. 255. If two circles are tangent to each other, the chords formed by a straight line drawn through the point of contact have the same ratio as the diameters of the circles.

Proposition XXV. Theorem.

365. *If two polygons are similar, they are composed of the same number of triangles, similar each to each, and similarly placed.*

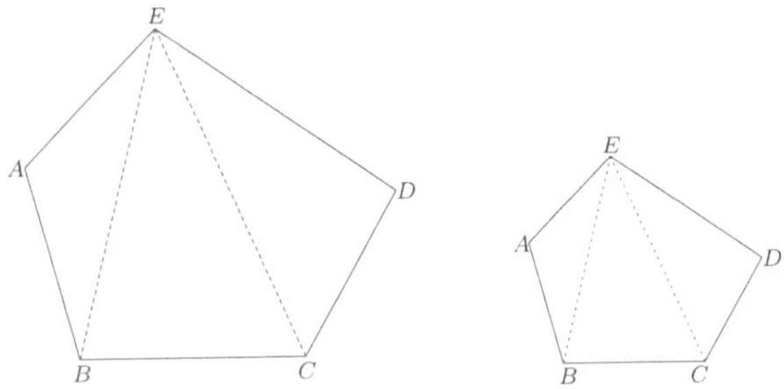

Let the polygons $ABCDE$ and $A'B'C'D'E'$ be similar.

From two homologous vertices, as E and E', draw diagonals EB, EC, and $E'B'$, $E'C'$.

To prove that the $_sEAB$, EBC, ECD *are similar, respectively, to the* $_sE'A'B'$, $E'B'C'$, $E'C'D'$.

Proof. The $_sEAB$ and $E'A'B'$ are similar. § 357

For

$$A = A',$$ § 351

and

$$AE : A'E' = AB : A'B'.$$ § 351

Now

$$ABC = A'B'C',$$ § 351

and

$$ABE = A'B'E'.$$ § 351

By subtracting,

$$EBC = E'B'C'.$$ Ax. 3

Now

$$EB : E'B' = AB : A'B'$$ § 351

and

$$BC : B'C' = AB : A'B'$$ § 351

$$EB : E'B' = BC : B'C'.$$ Ax. 1

$_sEBC$ and $E'B'C'$ are similar. § 357

In like manner $_sECD$ and $E'C'D'$ are similar. Q.E.D.

Proposition XXVI. Theorem.

366. Conversely: *If two polygons are composed of the same number of triangles, similar each to each, and similarly placed, the polygons are similar.*

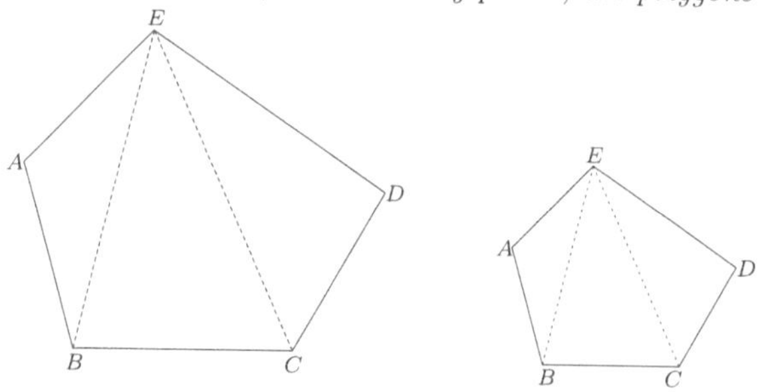

In the two polygons $ABCDE$ and $A'B'C'D'E'$, let the triangles AEB, BEC, CED be similar, respectively, to the triangles $A'E'B'$, $B'E'C'$, $C'E'D'$; and similarly placed.

To prove that $ABCDE$ is similar to $A'B'C'D'E'$.

Proof.

$$A = A' \qquad \text{§ 351}$$

Also,

$$ABE = A'B'E',$$

and

$$EBC = E'B'C'. \qquad \text{§ 351}$$

By adding,

$$ABC = A'B'C'. \qquad \text{Ax. 2}$$

In like manner, $BCD = B'C'D'$, $CDE = C'D'E'$, etc.

Hence, the polygons are mutually equiangular.

Also, $\dfrac{AB}{A'B'} = \left(\dfrac{EB}{E'B'}\right) = \dfrac{BC}{B'C'} = \left(\dfrac{EC}{E'C'}\right) = \dfrac{CD}{C'D'}$, etc. § 351

Hence, the polygons have their homologous sides proportional.

Therefore, the polygons are similar. § 351

Q.E.D.

THEOREMS.

Ex. 256. If two circles are tangent to each other externally, the corresponding segments of two lines drawn through the point of contact and terminated by the circumferences are proportional.

Ex. 257. In a parallelogram $ABCD$, a line DE is drawn, meeting the diagonal AC in F, the side BC in G, and the side AB produced in E. Prove that $\overline{DF}^2 = FG \times FE$.

Ex. 258. Two altitudes of a triangle are inversely proportional to the corresponding bases.

Ex. 259. Two circles touch at P. Through P three lines are drawn, meeting one circle in A, B, C, and the other in A', B', C'', respectively. Prove that the triangles ABC, $A'B'C''$ are similar.

Ex. 260. Two chords AB, CD intersect at M, and A is the middle point of the arc CD. Prove that the product $AB \times AM$ is constant if the chord AB is made to turn about the fixed point A.

Draw the diameter AE, and draw BE.

Ex. 261. If two circles touch each other, their common external tangent is the mean proportional between their diameters.

Let AB be the common tangent. Draw the diameters AC, BD. Join the point of contact P to A, B, C, and D. Show that APD and BPC are straight lines to each other, and that ${}_sCAB$, ABD are similar.

Ex. 262. If two circles are tangent internally, all chords of the greater circle drawn from the point of contact are divided proportionally by the circumference of the smaller circle.

Draw any two of the chords, and join the points where they meet the circumferences. The ${}_s$ thus formed are similar (Ex. 120).

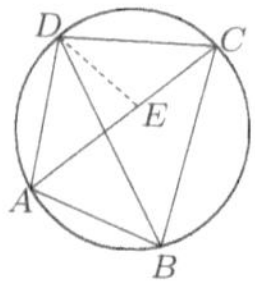

Ex. 263. In an inscribed quadrilateral, the product of the diagonals is equal to the sum of the products of the opposite sides.

Draw DE, making $CDE =$ ADB. The $_sABD$ and ECD are similar; and the $_sBCD$ and AED are similar.

Ex. 264. Two isosceles triangles with equal vertical angles are similar.

Ex. 265. The bisector of the vertical angle A of the triangle ABC intersects the base at D and the circumference of the circumscribed circle at E.

Show that $AB \times AC = AD \times AE$.

NUMERICAL PROPERTIES OF FIGURES.

PROPOSITION XXVII. THEOREM.

367. *If in a right triangle a perpendicular is drawn from the vertex of the right angle to the hypotenuse:*

1. *The triangles thus formed are similar to the given triangle, and to each other.*
2. *The perpendicular is the mean proportional between the segments of the hypotenuse.*
3. *Each leg of the right triangle is the mean proportional between the hypotenuse and its adjacent segment.*

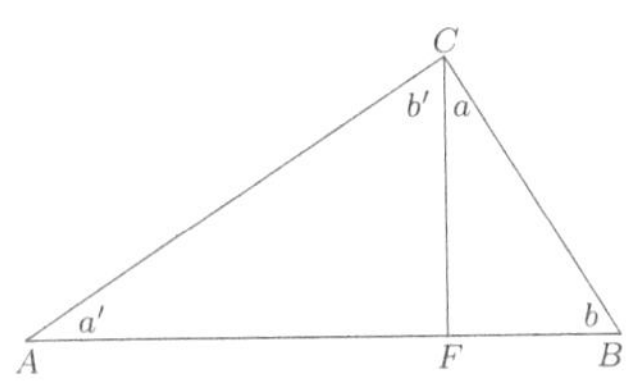

In the right triangle ABC, let CF be drawn from the vertex of the right angle C, perpendicular to AB.

1. *To prove that 's BCA, CFA, BFC are similar.*

Proof. The rt. $_sCFA$ and BCA are similar, § 356

since the a' is common.

The rt. $_sBFC$ and BCA are similar, § 356

since the b is common.

Since the $_sCFA$ and BFC are each similar to BCA, they are similar to each other. § 354

2. *To prove that*

$$AF : CF = CF : FB.$$

Proof. In the similar $_sCFA$ and BFC,

$$AF : CF = CF : FB. \quad \text{§ 351}$$

3. *To prove that*

$$AB : AC = AC : AF,$$

and

$$AB : BC = BC : BF.$$

Proof. In the similar $_sBCA$ and CFA,

$$AB : AC = AC : AF$$ § 351

In the similar $_sBCA$ and BFC,

$$AB : BC = BC : BF.$$ § 351

Q.E.D.

368. Cor. 1. *The squares of the two legs of a right triangle are proportional to the adjacent segments of the hypotenuse.*

From the proportions in § 367,3,

$$\overline{AC}^2 = AB \times AF, \text{ and } \overline{BC}^2 = AB \times BF.$$ § 327

Hence,

$$\frac{\overline{AC}^2}{\overline{BC}^2} = \frac{AB \times AF}{AB \times BF} = \frac{AF}{BF}.$$

369. Cor. 2. *The squares of the hypotenuse and either leg are proportional to the hypotenuse and the adjacent segment.*

For

$$\frac{\overline{AB}^2}{\overline{AC}^2} = \frac{AB \times AB}{AB \times AF} = \frac{AB}{AF}.$$

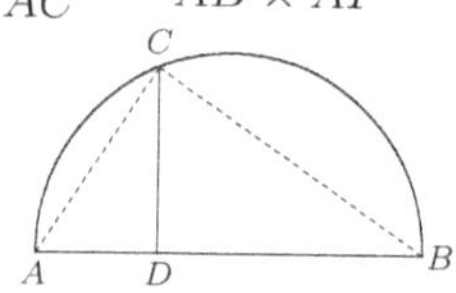

370. Cor. 3. *The perpendicular from any point in the circumference to the diameter of a circle is the mean proportional between the segments of the diameter.*

The chord drawn from any point in the circumference to either extremity of the diameter is the mean proportional between the diameter and the adjacent segment.

For

the ACB is a rt. . § 290

Proposition XXVIII. Theorem.

371. *The sum of the squares of the two legs of a right triangle is equal to the square of the hypotenuse.*

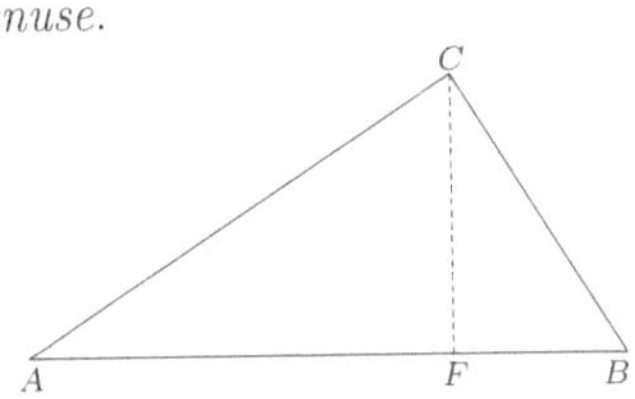

Let ABC be a right triangle with its right angle at C.

To prove that

$$\overline{AC}^2 + \overline{CB}^2 = \overline{AB}^2.$$

Proof.

Draw CF to AB.

Then

$$\overline{AC}^2 = AB \times AF,$$

and

$$\overline{CB}^2 = AB \times BF. \qquad \text{§ 367}$$

By adding,

$$\overline{AC}^2 + \overline{CB}^2 = AB(AF + BF) = \overline{AB}^2 \qquad \text{Q.E.D.}$$

372. Cor. 1. *The square of either leg of a right triangle is equal to the difference of the square of the hypotenuse and the square of the other leg.*

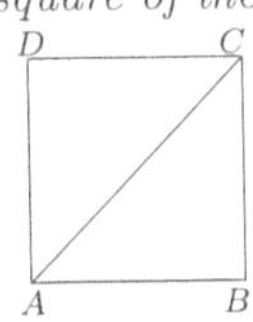

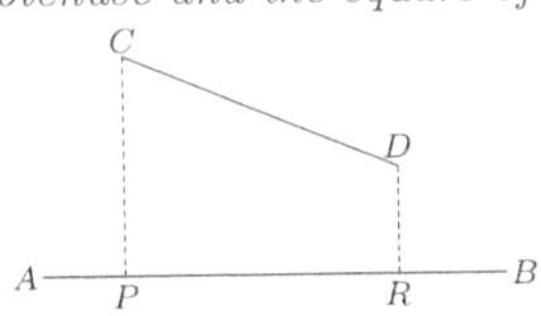

373. Cor. 2. *The diagonal and a side of a square are incommensurable.*

For

$$\overline{AC}^2 = \overline{AB}^2 + \overline{BC}^2 = 2\overline{AB}^2.$$

$$AC = AB\sqrt{2}.$$

374. Def. The **projection** of any line upon a second line is the segment of the second line included between the perpendiculars drawn to it from the extremities of the first line. Thus, PR is the projection of CD upon AB.

PROPOSITION XXIX. THEOREM.

375. *In any triangle, the square of the side opposite an acute angle is equal to the sum of the squares of the other two sides diminished by twice the product of one of those sides by the projection of the other upon that side.*

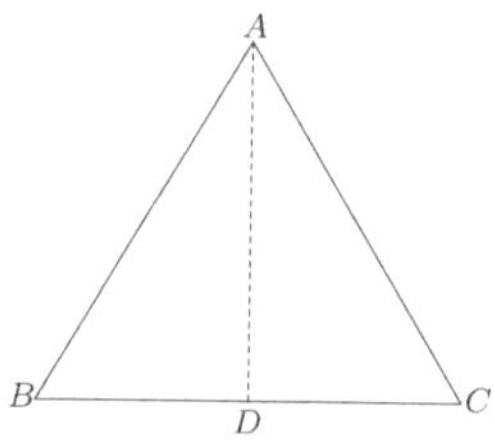

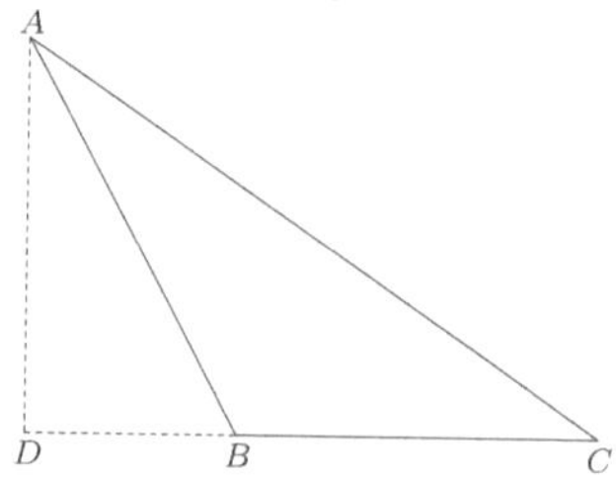

Let C be an acute angle of the triangle ABC, and DC the projection of AC upon BC.

To prove that $\overline{AB}^2 = \overline{BC}^2 + \overline{AC}^2 - 2BC \times DC$.

Proof. If D falls upon the base (Fig. 1),

$$DB = BC - DC.$$

If D falls upon the base produced (Fig. 2),

$$DB = DC - BC.$$

In either case,

$$\overline{DB}^2 = \overline{BC}^2 + \overline{DC}^2 - 2BC \times DC.$$

Add $\overline{AD}^2$ to both sides of this equality, and we have

$$\overline{AD}^2 + \overline{DB}^2 = \overline{BC}^2 + \overline{AD}^2 + \overline{DC}^2 - 2BC \times DC.$$

But

$$\overline{AD}^2 + \overline{DB}^2 = \overline{AB}^2$$

and

$$\overline{AD}^2 + \overline{DC}^2 = \overline{AC}^2 \qquad \S\ 371$$

Put $\overline{AB}^2$ and $\overline{AC}^2$ for their equals in the above equality.

Then

$$\overline{AB}^2 = \overline{BC}^2 + \overline{AC}^2 - 2BC \times DC. \qquad \text{Q.E.D.}$$

Proposition XXX. Theorem.

376. *In any obtuse triangle, the square of the side opposite the obtuse angle is equal to the sum of the squares of the other two sides increased by twice the product of one of those sides by the projection of the other upon that side.*

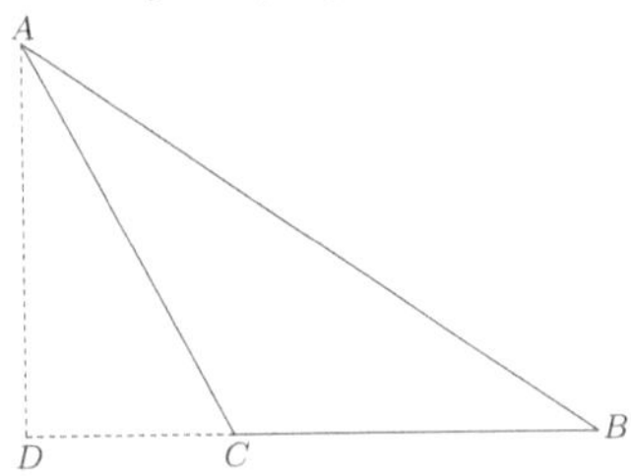

Let C be the obtuse angle of the triangle ABC, and CD be the projection of AC upon BC produced.

To prove that $\overline{AB}^2 = \overline{BC}^2 + \overline{AC}^2 + 2BC \times DC$.

Proof.

$$DB = BC + DC.$$

Squaring,

$$\overline{DB}^2 = \overline{BC}^2 + \overline{DC}^2 + 2BC \times DC.$$

Add $\overline{AD}^2$ to both sides, and we have

$$\overline{AD}^2 + \overline{DB}^2 = \overline{BC}^2 + \overline{AD}^2 + \overline{DC}^2 + 2BC \times DC.$$

But

$$\overline{AD}^2 + \overline{DB}^2 = \overline{AB}^2, \text{ and } \overline{AD}^2 + \overline{DC}^2 = \overline{AC}^2. \qquad \S\ 371$$

Put $\overline{AB}^2$ and $\overline{AC}^2$ for their equals in the above equality.

Then

$$\overline{AB}^2 = \overline{BC}^2 + \overline{AC}^2 + 2BC \times DC. \qquad \text{Q.E.D.}$$

Note 1. By the Principle of Continuity the last three theorems may be included in one theorem. Let the student explain.

Note 2. The last three theorems enable us to compute the lengths of the altitudes of a triangle if the lengths of the three sides are known.

Proposition XXXI. Theorem.

377. 1. *The sum of the squares of two sides of a triangle is equal to twice the square of half the third side increased by twice the square of the median upon that side.*

2. *The difference of the squares of two sides of a triangle is equal to twice the product of the third side by the projection of the median upon that side.*

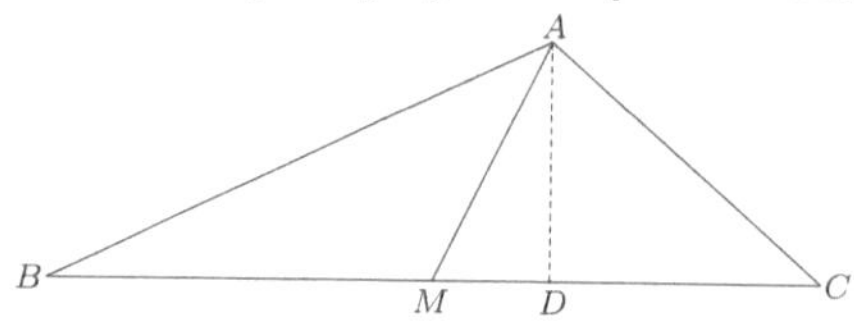

In the triangle ABC, let AM be the median and MD the projection of AM upon the side BC. Also, let AB be greater than AC.

To prove that

$$1.\ \overline{AB}^2 + \overline{AC}^2 = 2\overline{BM}^2 + 2\overline{AM}^2.$$

$$2.\ \overline{AB}^2 - \overline{AC}^2 = 2BC \times MD.$$

Proof. Since $AB > AC$, the AMB will be obtuse, and the AMC will be acute. § 155

Then

$$\overline{AB}^2 = \overline{BM}^2 + \overline{AM}^2 + 2BM \times MD, \qquad \text{§ 376}$$

and

$$\overline{AC}^2 = \overline{MC}^2 + \overline{AM}^2 - 2MC \times MD. \qquad \text{§ 375}$$

Add these two equalities, and observe that $BM = MC$.

Then

$$\overline{AB}^2 + \overline{AC}^2 = 2\overline{BM}^2 + 2\overline{AM}^2.$$

Subtract the second equality from the first.

Then

$$\overline{AB}^2 - \overline{AC}^2 = 2BC \times MD.$$

Q.E.D.

Note. This theorem enables us to compute the lengths of the medians of a triangle if the lengths of the three sides are known.

Proposition XXXII. Theorem.

378. *If two chords intersect in a circle, the product of the segments of one is equal to the product of the segments of the other.*

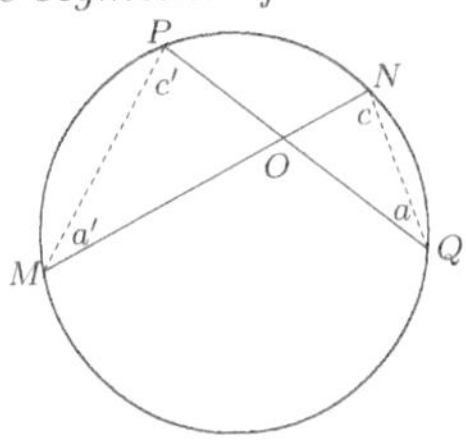

Let any two chords MN and PQ intersect at O.

To prove that

$$OM \times ON = OQ \times OP.$$

Proof.

Draw MP and NQ.

$a = a'$, § 289

(*each being measured by* $\frac{1}{2}$ arc PN).

And

$c = c'$, § 289

(*each being measured by* $\frac{1}{2}$ arc MQ).

the ${}_s NOQ$ and POM are similar. § 355

$OQ : OM = ON : OP.$ § 351

$OM \times ON = OQ \times OP.$ § 327

Q.E.D.

379. Scholium. This proportion may be written

$$\frac{OM}{OQ} = \frac{OP}{ON}, \text{ or } \frac{OM}{OQ} = \frac{1}{\frac{ON}{OP}};$$

that is, the ratio of two corresponding segments is equal to the *reciprocal* of the ratio of the other two segments.

Hence, these segments are said to be *reciprocally proportional.*

380. DEF. **A secant from a point to a circle** is understood to mean the segment of the secant lying between the point and the *second point* of intersection of the secant and circumference.

PROPOSITION XXXIII. THEOREM.

381. *If from a point without a circle a secant and a tangent are drawn, the tangent is the mean proportional between the whole secant and its external segment.*

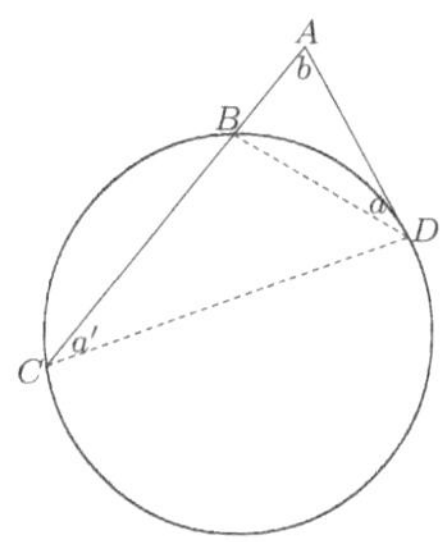

Let AD be a tangent and AC a secant drawn from the point A to the circle BCD.

To prove that $AC : AD = AD : AB$.

Proof.

Draw DC and DB.

The $_{s}ADC$ and ABD are similar. § 355

For b is common; and $a' = a$, §§ 289, 295
(*each being measured by* $\frac{1}{2}$ arc BD).

$$AC : AD = AD : AB.$$ § 351

Q.E.D.

382. COR. *If from a fixed point without a circle a secant is drawn, the product of the secant and its external segment is constant in whatever direction the secant is drawn.*

For

$$AC \times AB = \overline{AD}^2.$$ § 327

PROPOSITION XXXIV. THEOREM.

383. *The square of the bisector of an angle of a triangle is equal to the product of the sides of this angle diminished by the product of the segments made by the bisector upon the third side of the triangle.*

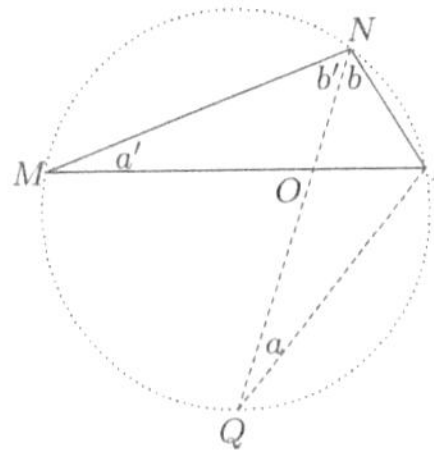

Let NO bisect the angle MNP of the triangle MNP.

To prove that $\overline{NO}^2 = NM \times NP - OM \times OP$.

Proof.

Circumscribe the $\odot MNP$ about the MNP. § 314

Produce NO to meet the circumference in Q, and draw PQ.

The ${}_s NQP$ and NMO are similar. § 355

For

$$b = b'$$ Hyp.

and

$$a = a'$$ § 289

Whence

$$NQ : NM = NP : NO.$$ § 351

$$NM \times NP = NQ \times NO$$

$$= (NO + OQ)NO$$

$$= \overline{NO}^2 + NO \times OQ.$$

But

$$NO \times OQ = MO \times OP.$$ § 378

$$MN \times NP = \overline{NO}^2 + MO \times OP.$$

Whence

$$\overline{NO}^2 = NM \times NP = MO \times OP.$$ Ax. 3

Q.E.D.

NOTE. This theorem enables us to compute the lengths of the bisectors of the angles of a triangle if the lengths of the sides are known.

PROPOSITION XXXV. THEOREM.

384. *In any triangle the product of two sides is equal to the product of the diameter of the circumscribed circle by the altitude upon the third side.*

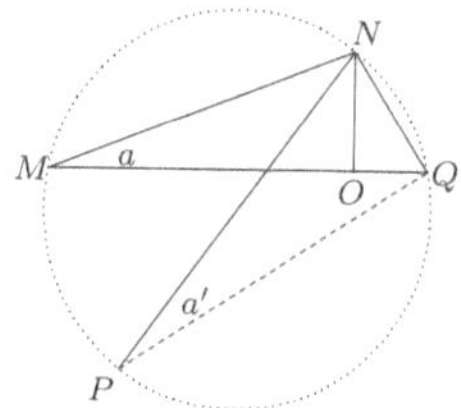

Let NMQ be a triangle, NO the altitude, and QNMP the circle circumscribed about the triangle NMQ.

Draw the diameter NP, and draw PQ.

To prove that $NM \times NQ = NP \times NO$.

Proof. In the $_sNOM$ and NQP,

NOM is a rt. , Hyp.

NQP is a rt. , § 290

and $a = a'$, § 289

(*each being measured by* $\frac{1}{2}$ arc NQ).

$_sNOM$ and NQP are similar. § 356

Whence

$NM : NP = NO : NQ$. § 351

$NM \times NQ = NP \times NO$. § 327

Q.E.D.

NOTE. This theorem enables us to compute the length of the radius of a circle circumscribed about a triangle, if the lengths of the three sides of the triangle are known.

Ex. 266. If OE, OF, OG are the perpendiculars from any point O within the triangle ABC upon the sides AB, BC, CA, respectively, show that $\overline{AE}^2 + \overline{BF}^2 + \overline{CG}^2 = \overline{EB}^2 + \overline{FC}^2 + \overline{GA}^2$.

THEOREMS.

Ex. 267. The sum of the squares of the segments of two perpendicular chords is equal to the square of the diameter of the circle.

If AB, CD are the chords, draw the diameter BE, draw AC, ED, BD. Prove that $AC = ED$, and apply § 371.

Ex. 268. The tangents to two intersecting circles drawn from any point in their common chord produced, are equal. (§ 381.)

Ex. 269. The common chord of two intersecting circles, if produced, will bisect their common tangents. (§ 381.)

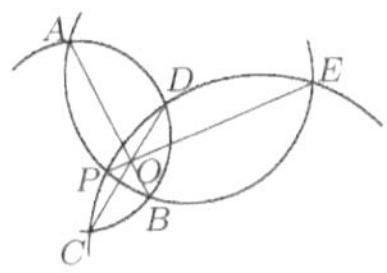

Ex. 270. If three circles intersect one another, the common chords all pass through the same point.

Let two of the chords AB and CD meet at O. Join the point of intersection E to O, and suppose that EO produced meets the same two circles at two different points P and Q. Then prove that $OP = OQ$ (§ 378); hence, that the points P and Q coincide.

Ex. 271. If two circles are tangent to each other, the common internal tangent bisects the two common external tangents.

Ex. 272. If the perpendiculars from the vertices of the triangle ABC upon the opposite sides intersect at D, show that

$$\overline{AB}^2 - \overline{AC}^2 = \overline{BD}^2 - \overline{CD}^2.$$

Ex. 273. In an isosceles triangle, the square of a leg is equal to the square of any line drawn from the vertex to the base, increased by the product of the segments of the base.

Ex. 274. The squares of two chords drawn from the same point in a circumference have the same ratio as the projections of the chords on the diameter drawn from the same point.

Ex. 275. The difference of the squares of two sides of a triangle is equal to the difference of the squares of the segments of the third side, made by the perpendicular on the third side from the opposite vertex.

Ex. 276. E is the middle point of BC, one of the parallel sides of the trapezoid $ABCD$; AE and DE produced meet DC and AB produced at F and G, respectively. Show that FG is parallel to DA.

$_sAGD$ and BGE are similar; and $_sAFD$ and EFC are similar.

Ex. 277. If two tangents are drawn to a circle at the extremities of a diameter, the portion of a third tangent intercepted between them is divided at its point of contact into segments whose product is equal to the square of the radius.

Ex. 278. If two exterior angles of a triangle are bisected, the line drawn from the point of intersection of the bisectors to the opposite angle of the triangle bisects that angle.

Ex. 279. The sum of the squares of the diagonals of a quadrilateral is equal to twice the sum of the squares of the lines that join the middle points of the opposite sides.

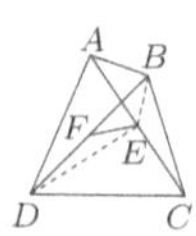

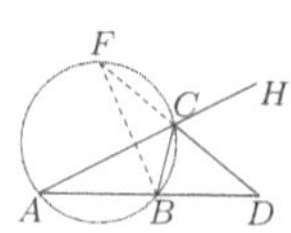

Ex. 280. The sum of the squares of the four sides of any quadrilateral is equal to the sum of the squares of the diagonals, increased by four times the square of the line joining the middle points of the diagonals.

Apply § 377 to the $_s$ ABC and ADC, add the results, and eliminate $\overline{BE}^2 + \overline{DE}^2$ by applying § 377 to the BDE.

Ex. 281. The square of the bisector of an exterior angle of a triangle is equal to the product of the external segments determined by the bisector upon one of the sides, diminished by the product of the other two sides.

Let CD bisect the exterior BCH of the ABC. $_s$ ACD and FCB are similar (§ 355). Apply § 382.

Ex. 282. If a point O is joined to the vertices of a triangle ABC; through any point A' in OA a line parallel to AB is drawn, meeting OB at B'; through B' a line parallel to BC, meeting OC at C'; and C' is joined to A'; the triangle $A'B'C'$ is similar to the triangle ABC.

Ex. 283. If the line of centres of two circles meets the circumferences at the consecutive points A, B, C, D, and meets the common external tangent at P, then $PA \times PD = PB \times PC$.

Ex. 284. The line of centres of two circles meets the common external tangent at P, and a secant is drawn from P, cutting the circles at the consecutive points E, F, G, H. Prove that $PE \times PH = PF \times PG$.

Draw radii to the points of contact, and to E, F, G, H. Let fall $_s$ on PH from the centres of the $\odot_s$. The various pairs of $_s$ are similar.

Ex. 285. If a line drawn from a vertex of a triangle divides the opposite side into segments proportional to the adjacent sides, the line bisects the angle at the vertex.

PROBLEMS OF CONSTRUCTION.

Proposition XXXVI. Problem.

385. *To divide a given straight line into parts proportional to any number of given lines.*

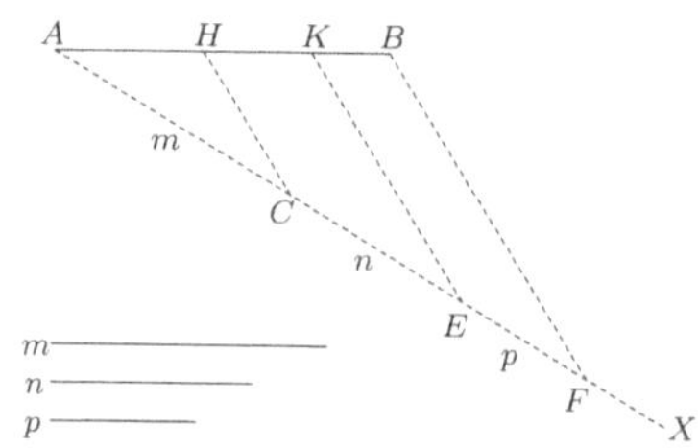

Let AB, m, n, and p be given straight lines.

To divide AB into parts proportional to m, n, and p.

Draw AX, making any convenient with AB.

On AX take AC equal to m, CE to n, EF to p.

Draw BF.

From E and C draw EK and CH $\parallel$ to FB.

Through A draw a line $\parallel$ to BF.

K and H are the division points required.

Proof.

$$\frac{AH}{AC} = \frac{HK}{CE} = \frac{KB}{EF}, \qquad \S\ 344$$

(*if two lines are cut by any number of parallels, the corresponding intercepts are proportional*).

Substitute m, n, and p for their equals AC, CE, and EF.

Then

$$\frac{AH}{m} = \frac{HK}{n} = \frac{KB}{p}. \qquad \text{Q.E.F.}$$

Ex. 286. Divide a line 12 inches long into three parts proportional to the numbers 3, 5, 7.

Proposition XXXVII. Problem.

386. *To find the fourth proportional to three given straight lines.*

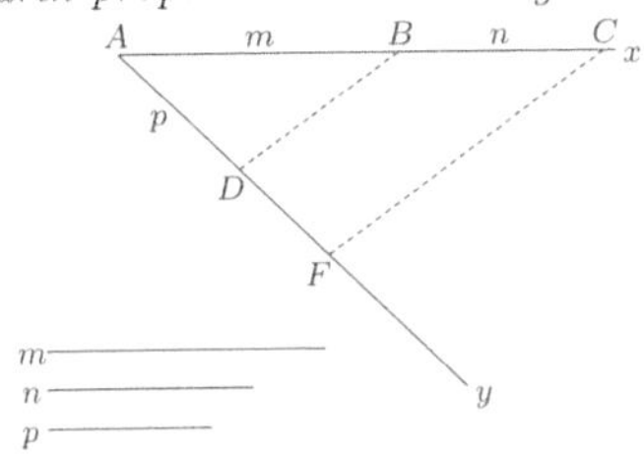

Let the three given lines be m, n, and p.

To find the fourth proportional to m, n, and p.

Draw Ax and Ay containing any convenient angle.

On Ax take AB equal to m, BC to n.

On Ay take AD equal to p.

Draw BD.

From C draw $CF \parallel$ to BD, meeting Ay at F.

DF is the fourth proportional required.

Proof.

$$AB : BC = AD : DF, \qquad \S\ 342$$

(*a line drawn through two sides of a* $\parallel$ *to the third side divides those sides proportionally*).

Substitute m, n, and p for their equals AB, BC, and AD.

Then

$$m : n = p : DF \qquad \text{Q.E.F.}$$

Ex. 287. The square of the altitude of an equilateral triangle is equal to three fourths of the square of one side of the triangle.

PROPOSITION XXXVIII. PROBLEM.

387. *To find the third proportional to two given straight lines.*

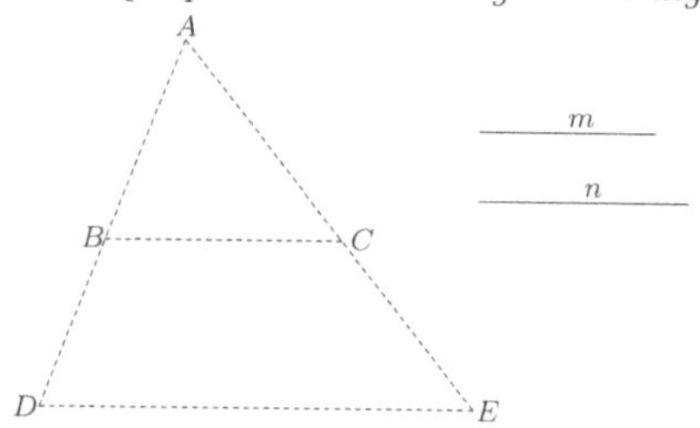

Let m and n be the two given straight lines.

To find the third proportional to m and n.

Construct any convenient angle A,

and take AB equal to m, AC equal to n.

Produce AB to D, making BD equal to AC.

Draw BC.

Through D draw $DE \parallel$ to BC, meeting AC produced at E.

CE is the third proportional required.

Proof.

$$AB : BD = AC : CE, \qquad \S\ 342$$

(*a line drawn through two sides of a parallel to the third side divides those sides proportionally*).

Substitute, in the above proportion, AC for its equal BD.

Then

$$AB : AC = AC : CE,$$

that is,

$$m : n = n : CE. \qquad \text{Q.E.F.}$$

Ex. 288. Construct x, if (1) $x = \dfrac{ab}{c}$, (2) $x = \dfrac{a^2}{c}$.

Special cases: (1) $a = 2$, $b = 8$, $c = 4$; (2) $a = 3$, $b = 7$, $c = 11$; (3) $a = 2$, $c = 3$; (4) $a = 3$, $c = 5$; (5) $a = 2c$.

PROPOSITION XXXIX. PROBLEM.

388. *To find the mean proportional between two given straight lines.*

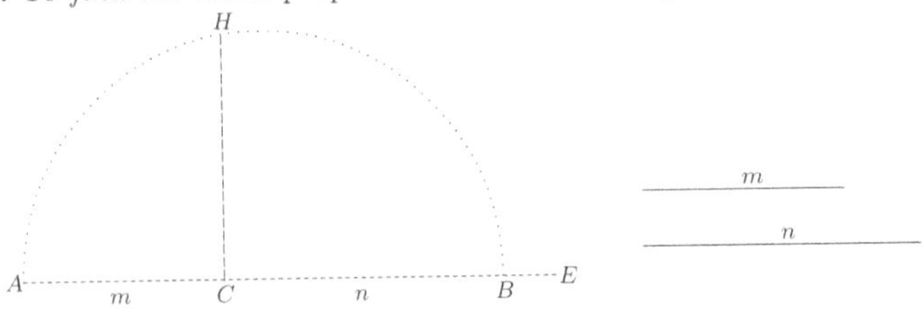

Let the two given lines be m and n.

To find the mean proportional between m and n.

On the straight line AE

take AC equal to m, and CB equal to n.

On AB as a diameter describe a semicircumference.

At C erect the CH meeting the circumference at H.

CH is the mean proportional between m and n.

Proof.

$$AC : CH = CH : CB$$ § 370

(*the let fall from a point in a circumference to the diameter of a circle is the mean proportional between the segments of the diameter*).

Substitute for AC and CB their equals m and n.

Then

$$m : CH = CH : n.$$ Q.E.F.

389. DEF. A straight line is divided **in extreme and mean ratio**, when one of the segments is the mean proportional between the whole line and the other segment.

Ex. 289. Construct x, if $x = \sqrt{ab}$.

Special cases: (1) $a = 2$, $b = 3$; (2) $a = 1$, $b = 6$; (3) $a = 3$, $b = 7$.

Proposition XL. Problem.

390. *To divide a given line in extreme and mean ratio.*

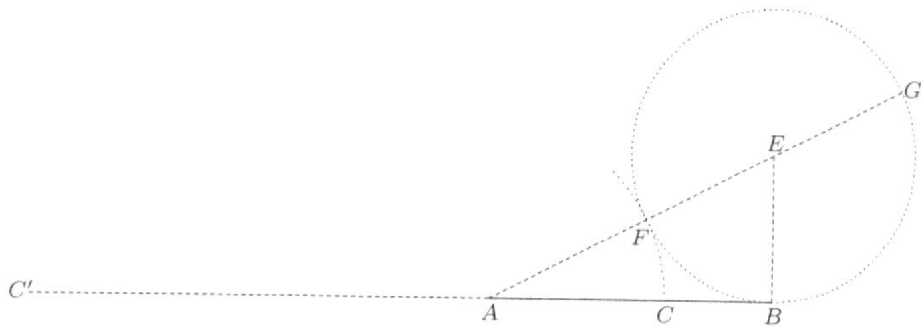

Let AB be the given line.

To divide AB in extreme and mean ratio.

At B erect a BE equal to half of AB.

From E as a centre, with a radius equal to EB, describe a ⊙.

Draw AE, meeting the circumference in F and G.

On AB take AC equal to AF.

On BA produced take AC' equal to AG.

Then AB is divided internally at C and externally at C' in extreme and mean ratio.

$$AG : AB = AB : AF. \qquad \S\ 381$$

$$\begin{aligned}
\overline{AB}^2 &= AF \times AG\\
&= AC(AF + AG)\\
&= AC(AC + AB)\\
&= \overline{AC}^2 + AB \times AC.\\
\overline{AB}^2 - AB \times AC &= \overline{AC}^2.\\
AB(AB - AC) &= \overline{AC}^2.\\
AB \times CB &= \overline{AC}^2.
\end{aligned}$$

$$\begin{aligned}
\overline{AB}^2 &= AG \times AF\\
&= C'A(AG - AF)\\
&= C'A(C'A - AB)\\
&= \overline{C'A}^2 - AB \times C'A.\\
\overline{AB}^2 + AB \times C'A &= \overline{C'A}^2.\\
AB(AB + C'A) &= \overline{C'A}^2.\\
AB \times C'B &= \overline{C'A}^2.
\end{aligned}$$

Q.E.F.

Proposition XLI. Problem.

391. *Upon a given line homologous to a given side of a given polygon, to construct a polygon similar to the given polygon.*

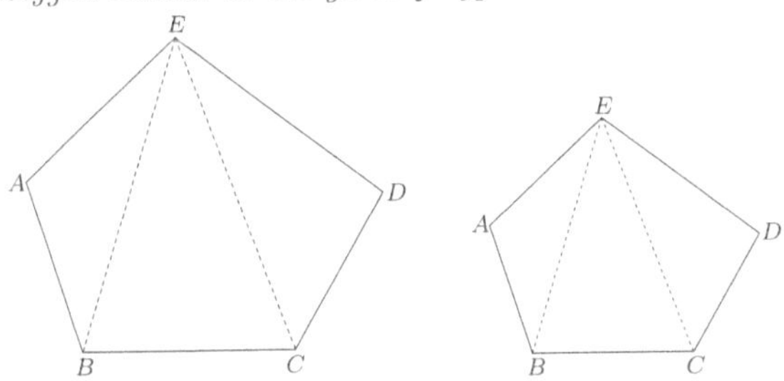

Let $A'E'$ be the given line homologous to AE of the given polygon $ABCDE$.

To construct on $A'E'$ a polygon similar to the given polygon.

From E draw the diagonals EB and EC.

From E' draw $E'B'$, $E'C'$, and $E'D'$,

making ’s $A'E'B'$, $B'E'C'$, and $C'E'D'$ equal, respectively, to

$_s AEB$, BEC, and CED.

From A' draw $A'B'$, making $E'A'B'$ equal to EAB,

and meeting $E'B'$ at B'.

From B' draw $B'C'$, making $E'B'C'$ equal to EBC,

and meeting $E'C'$ at C'.

From C' draw $C'D'$, making $E'C'D'$ equal to ECD,

and meeting $E'D'$ at D'.

Then $A'B'C'D'E'$ is the required polygon.

Proof.

The $_s ABE$, $A'B'E'$, etc., are similar. § 354

Therefore, the two polygons are similar. § 366

Q.E.F.

PROBLEMS OF CONSTRUCTION.

Ex. 290. To divide one side of a given triangle into segments proportional to the adjacent sides (§ 348).

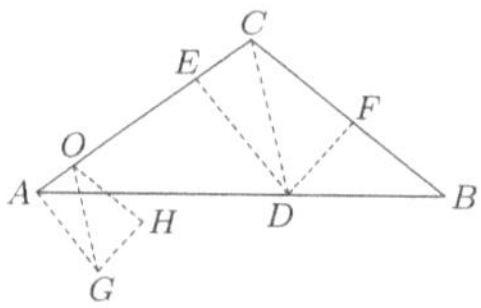

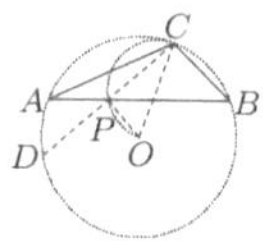

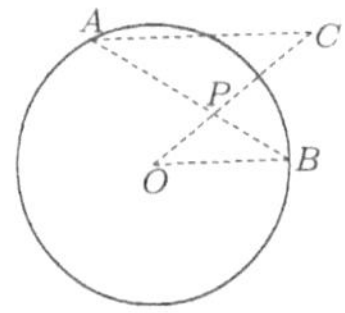

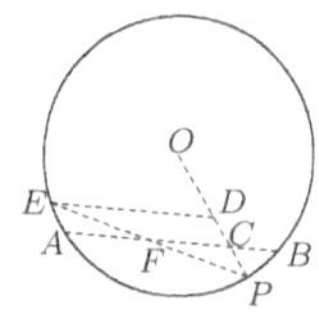

Ex. 291. To find in one side of a given triangle a point whose distances from the other sides shall be to each other in the given ratio $m : n$.

Take $AG = m$ to AC, $GH = n$ to BC. Draw $CD \parallel$ to OG.

Ex. 292. Given an obtuse triangle; to draw a line from the vertex of the obtuse angle to the opposite side which shall be the mean proportional between the segments of that side.

Ex. 293. Through a given point P within a given circle to draw a chord AB so that the ratio $AP : BP$ shall equal the given ratio $m : n$.

Draw OPC so that $OP : PC = n : m$. Draw CA equal to the fourth proportional to n, m, and the radius of the circle.

Ex. 294. To draw through a given point P in the arc subtended by a chord AB a chord which shall be bisected by AB.

On radius OP take CD equal to CP. Draw $DE \parallel$ to BA.

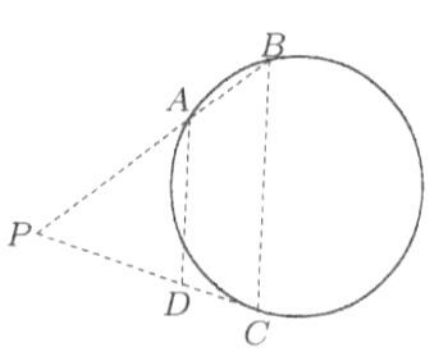

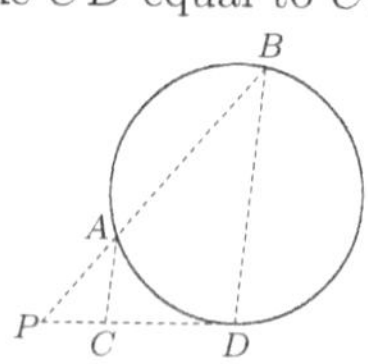

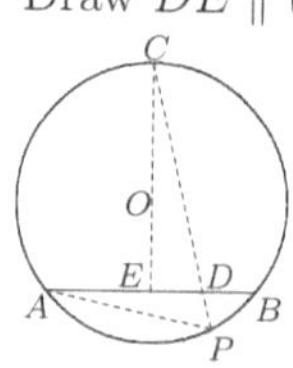

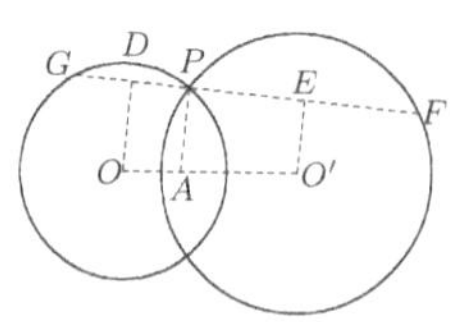

Ex. 295. To draw through a given external point P a secant PAB to a given circle so that the ratio $PA : AB$ shall equal the given ratio $m : n$.

$$PD : DC = m : n. \quad PD : PA = PA : PC.$$

Ex. 296. To draw through a given external point P a secant PAB to a given circle so that $\overline{AB}^2 = PA \times PB$.

$$PC : CD = CD : PD. \quad PA = CD.$$

Ex. 297. To find a point P in the arc subtended by a given chord AB so that the ratio $PA : PB$ shall equal the given ratio $m : n$.

Ex. 298. To draw through one of the points of intersection of two circles a secant so that the two chords that are formed shall be in the given ratio $m : n$.

Ex. 299. Having given the greater segment of a line divided in extreme and mean ratio, to construct the line.

Ex. 300. To construct a circle which shall pass through two given points and touch a given straight line.

Ex. 301. To construct a circle which shall pass through a given point and touch two given straight lines.

Ex. 302. To inscribe a square in a semicircle.

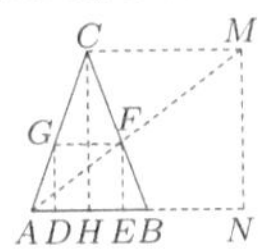

Ex. 303. To inscribe a square in a given triangle.

Let $DEFG$ be the required inscribed square. Draw $CM \parallel$ to AB, meeting AF produced in M. Draw CH and MN to AB, and produce AB to meet MN at N. The ${}_s ACM$, AGF are similar; also, the ${}_s AMN$, AFE are similar. By these triangles show that the figure $CMNH$ is a square. By constructing this square, the point F can be found.

Ex. 304. To inscribe in a given triangle a rectangle similar to a given rectangle.

Ex. 305. To inscribe in a circle a triangle similar to a given triangle.

Ex. 306. To inscribe in a given semicircle a rectangle similar to a given rectangle.

Ex. 307. To circumscribe about a circle a triangle similar to a given triangle.

Ex. 308. To construct the expression, $x = \frac{2abc}{de}$; that is, $\frac{2ab}{d} \times \frac{c}{e}$.

Ex. 309. To construct two straight lines, having given their sum and their ratio.

Ex. 310. To construct two straight lines, having given their difference and their ratio.

Ex. 311. Given two circles, with centres O and O', and a point A in their plane, to draw through the point A a straight line, meeting the circumferences at B and C, so that $AB : AC = m : n$.

PROBLEMS OF COMPUTATION.

Ex. 312. To compute the altitudes of a triangle in terms of its sides.

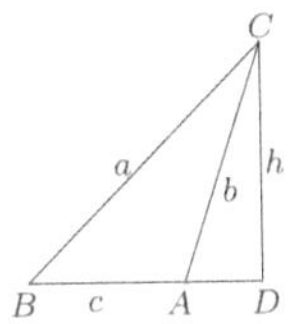

At least one of the angles A or B is acute. Suppose B is acute.

In the CDB,

$$h^2 = a^2 - \overline{BD}^2, \qquad \text{§ 372}$$

In the ABC,

$$b^2 = a^2 + c^2 - 2c \times BD. \qquad \text{§ 376}$$

Whence

$$BD = \frac{a^2 + c^2 - b^2}{2c}.$$

Hence,

$$h^2 = a^2 - \frac{(a^2 + c^2 - b^2)^2}{4c^2} = \frac{4a^2c^2 - (a^2 + c^2 - b^2)^2}{4c^2}$$

$$= \frac{(2ac + a^2 + c^2 - b^2)(2ac - a^2 - c^2 + b^2)}{4c^2}$$

$$= \frac{\{(a + c)^2 - b^2\}\{b^2 - (a - c)^2\}}{4c^2}$$

$$= \frac{(a + b + c)(a + c - b)(b + a - c)(b - a + c)}{4c^2}.$$

Let

$$a + b + c = 2s.$$

Then

$$a + c - b = 2(s - b),$$

$$b + a - c = 2(s - c),$$

$$b - a + c = 2(s - a).$$

Hence,

$$h^2 = \frac{2s \times 2(s - a) \times 2(s - b) \times 2(2 - c)}{4c^2}.$$

By simplifying, and extracting the square root,

$$h = \frac{2}{c}\sqrt{s(s-a)(s-b)(s-c)}.$$

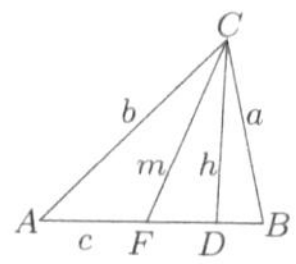

Ex. 313. To compute the medians of a triangle in terms of its sides. By § 377,

$$a^2 + b^2 = 2m^2 + 2\left(\frac{c}{2}\right)^2.$$

Whence

$$4m^2 = 2(a^2 + b^2) - c^2.$$

$$m = \frac{1}{2}\sqrt{2(a^2 + b^2) - c^2}.$$

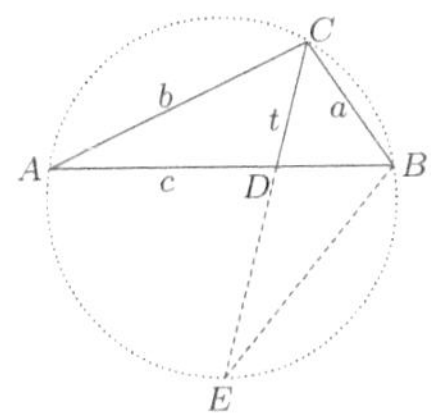

Ex. 314. To compute the bisectors of a triangle in terms of the sides.

By § 383,

$$t^2 = ab - AD \times BD.$$

By §348,

$$\frac{AD}{B} = \frac{BD}{a} = \frac{AD + BD}{a + b} = \frac{c}{a + b}.$$

$$AD = \frac{bc}{a + b}, \text{ and } BD = \frac{ac}{a + b}.$$

Whence

$$\begin{aligned} t^2 &= ab - \frac{abc^2}{(a + b)^2} \\ &= ab\left[1 - \frac{c^2}{(a + b)^2}\right] \\ &= \frac{ab\{(a + b)^2 - c^2\}}{(a + b)^2} \\ &= \frac{ab(a + b + c)(a + b - c)}{(a + b)^2} \\ &= \frac{ab \times 2s \times 2(s - c)}{(a + b^2)}. \end{aligned}$$

Whence

$$t = \frac{2}{a + b}\sqrt{abs(s - c)}.$$

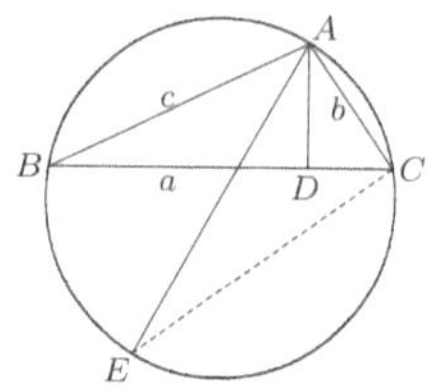

Ex. 315. To compute the radius of the circle circumscribed about a triangle in terms of the sides of the triangle.

By §384,

$$AC \times AB = AE \times AD,$$

or,

$$bc = 2B \times AD.$$

But

$$AD = \frac{2}{a}\sqrt{s(s-a)(s-b)(s-c)}. \qquad \text{Ex. 312}$$

$$R = \frac{abc}{4\sqrt{s(s-a)(s-b)(s-c)}}.$$

Ex. 316. If the sides of a triangle are 3, 4, and 5, is the angle opposite 5 right, acute, or obtuse?

Ex. 317. If the sides of a triangle are 7, 9, and 12, is the angle opposite 12 right, acute, or obtuse?

Ex. 318. If the sides of a triangle are 7, 9, and 11, is the angle opposite 11 right, acute, or obtuse?

Ex. 319. The legs of a right triangle are 8 inches and 12 inches; find the lengths of the projections of these legs upon the hypotenuse, and the distance of the vertex of the right angle from the hypotenuse.

Ex. 320. If the sides of a triangle are 6 inches, 9 inches, and 12 inches, find the lengths (1) of the altitudes; (2) of the medians; (3) of the bisectors; (4) of the radius of the circumscribed circle.

Ex. 321. A line is drawn parallel to a side AB of a triangle ABC, cutting AC in D, BC in E. If $AD : DC = 2 : 3$, and $AB = 20$ inches, find DE.

Ex. 322. The sides of a triangle are 9, 12, 15. Find the segments of the sides made by bisecting the angles.

Ex. 323. A tree casts a shadow 90 feet long, when a post 6 feet high casts a shadow 4 feet long. How high is the tree?

Ex. 324. The lower and upper bases of a trapezoid are a, b, respectively; and the altitude is h. Find the altitudes of the two triangles formed by producing the legs until they meet.

Ex. 325. The sides of a triangle are 6, 7, 8, respectively. In a similar triangle the side homologous to 8 is 40. Find the other two sides.

Ex. 326. The perimeters of two similar polygons are 200 feet and 300 feet. If a side of the first is 24 feet, find the homologous side of the second.

Ex. 327. How long a ladder is required to reach a window 24 feet high, if the lower end of the ladder is 10 feet from the side of the house?

Ex. 328. If the side of an equilateral triangle is a, find the altitude.

Ex. 329. If the altitude of an equilateral triangle is h, find the side.

Ex. 330. Find the length of the longest chord and of the shortest chord that can be drawn through a point 6 inches from the centre of a circle whose radius is 10 inches.

Ex. 331. The distance from the centre of a circle to a chord 10 feet long is 12 feet. Find the distance from the centre to a chord 24 feet long.

Ex. 332. The radius of a circle is 5 inches. Through a point 3 inches from the centre a diameter is drawn, and also a chord perpendicular to the diameter. Find the length of this chord, and the distance from one end of the chord to the ends of the diameter.

Ex. 333. The radius of a circle is 6 inches. Find the lengths of the tangents drawn from a point 10 inches from the centre, and also the length of the chord joining the points of contact.

Ex. 334. The sides of a triangle are 407 feet, 368 feet, and 351 feet. Find the three bisectors and the three altitudes.

Ex. 335. If a chord 8 inches long is 8 inches distant from the centre of the circle, find the radius, and the chords drawn from the end of the chord to the ends of the diameter which bisects the chord.

Ex. 336. From the end of a tangent 20 inches long a secant is drawn through the centre of the circle. If the external segment of this secant is 8 inches, find the radius of the circle.

Ex. 337. The radius of a circle is 13 inches. Through a point 5 inches from the centre any chord is drawn. What is the product of the two segments of the chord? What is the length of the shortest chord that can be drawn through the point?

Ex. 338. The radius of a circle is 9 inches and the length of a tangent 12 inches. Find the length of a line drawn from the extremity of the tangent to the centre of the circle.

Ex. 339. Two circles have radii of 8 inches and 3 inches, respectively, and the distance between their centres is 15 inches. Find the lengths of their common tangents.

Ex. 340. Find the segments of a line 10 inches long divided in extreme and mean ratio.

Ex. 341. The sides of a triangle are 4, 5, 5. Is the largest angle acute, right, or obtuse?

Ex. 342. Find the third proportional to two lines whose lengths are 28 feet and 42 feet.

Ex. 343. If the sides of a triangle are a, b, c, respectively, find the lengths of the three altitudes.

Ex. 344. The diameter of a circle is 30 feet and is divided into five equal parts. Find the lengths of the chords drawn through the points of division perpendicular to the diameter.

Ex. 345. The radius of a circle is 2 inches. From a point 4 inches from the centre a secant is drawn so that the internal segment is 1 inch. Find the length of the secant.

Ex. 346. The sides of a triangular pasture are 1551 yards, 2068 yards, 2585 yards. Find the median to the longest side.

Ex. 347. The diagonal of a rectangle is d, and the perimeter is p. Find the sides.

Ex. 348. The radius of a circle is r. Find the length of a chord whose distance from the centre is $\frac{1}{2}r$.

BOOK IV. AREAS OF POLYGONS.

392. Def. The **unit of surface** is a square whose side is a *unit of length.*

393. Def. The **area of a surface** is the *number of units of surface* it contains.

394. Def. Plane figures that *have equal areas but cannot be made to coincide* are called **equivalent**.

Note. In propositions relating to *areas*, the words "rectangle," "triangle," etc., are often used for "area of rectangle," "area of triangle," etc.

Proposition I. Theorem.

395. *Two rectangles having equal altitudes are to each other as their bases.*

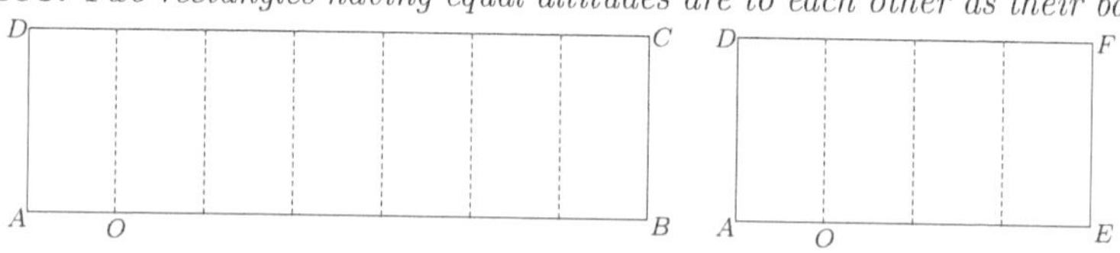

Let the rectangles AC and AF have the same altitude AD.

To prove that rect. AC : rect. AF = base AB : base AE.

Case 1. *When AB and AE are commensurable.*

Proof. Suppose AB and AE have a common measure, as AO, which is contained m times in AB and n times in AE.

Then

$$AB : AE = m : n.$$

Apply AO as a unit of measure to AB and AE, and at the several points of division erect $\perp$s.

The

$$\text{rect. } AC \text{ is divided into } m \text{ rectangles,}$$

and the

$$\text{rect. } AF \text{ is divided into } n \text{ rectangles.}$$ § 107

These rectangles are all equal. § 186

Hence,

$$\text{rect. } AC : \text{rect. } AF = m : n.$$

Therefore,

$$\text{rect. } AC : \text{rect. } AF = AB : AE.$$ Ax. 1

CASE 2. *When AB and AE are incommensurable.*

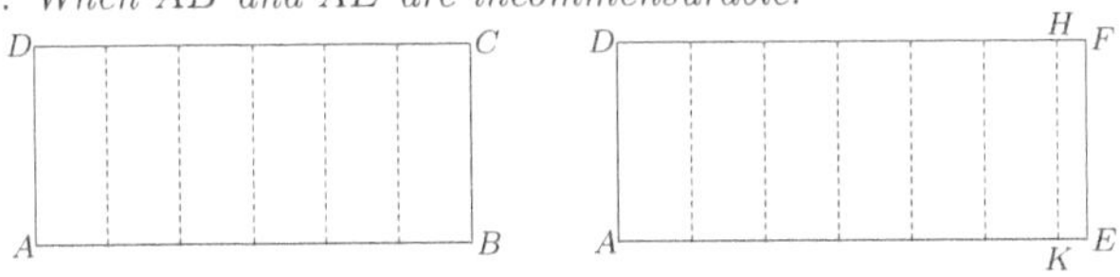

Proof. Divide AB into any number of equal parts, and apply one of them to AE as many times as AE will contain it.

Since AB and AE are incommensurable, a certain number of these parts will extend from A to some point K, leaving a remainder KE less than one of the equal parts of AB.

Draw $KH \parallel$ to EF.

Then AB and AK are commensurable by construction.

Therefore,

$$\frac{\text{rect. } AH}{\text{rect. } AC} = \frac{AK}{AB}.$$ Case 1

If the number of equal parts into which AB is divided is indefinitely increased, the varying values of these ratios will continue equal, and approach for their respective limits the ratios

$$\frac{\text{rect. } AF}{\text{rect. } AC} \text{ and } \frac{AE}{AB}. \text{ (See § 287.)}$$

$$\frac{\text{rect. } AF}{\text{rect. } AC} = \frac{AE}{AB}.$$ § 284

Q.E.D.

396. COR. *Two rectangles having equal bases are to each other as their altitudes.*

PROPOSITION II. THEOREM.

397. *Two rectangles are to each other as the products of their bases by their altitudes.*

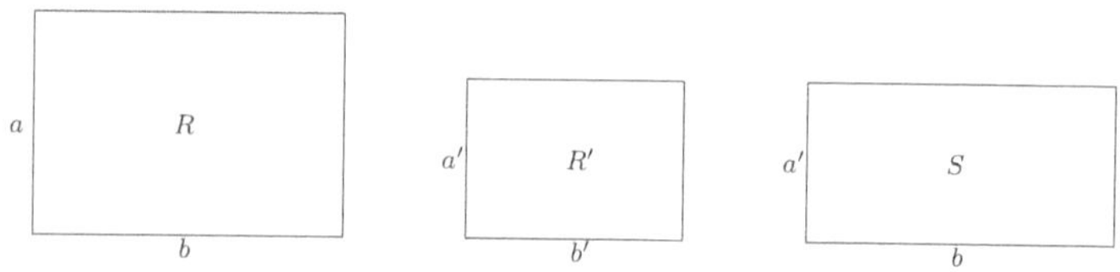

Let R and R' be two rectangles, having for their bases b and b', and for their altitudes a and a', respectively.

To prove that

$$\frac{R}{R'} = \frac{a \times b}{a' \times b'}.$$

Proof. Construct the rectangle S, with its base equal to that of R, and its altitude equal to that of R'.

Then

$$\frac{R}{S} = \frac{a}{a'}, \qquad \text{§ 396}$$

and

$$\frac{S}{R'} = \frac{b}{b'}. \qquad \text{§ 395}$$

The products of the corresponding members of these equations give

$$\frac{R}{R'} = \frac{a \times b}{a' \times b'}. \qquad \text{Q.E.D.}$$

Ex. 349. Find the ratio of a rectangular lawn 72 yards by 49 yards to a grass turf 18 inches by 14 inches.

Ex. 350. Find the ratio of a rectangular courtyard $18\frac{1}{2}$ yards by $15\frac{1}{2}$ yards to a flagstone 31 inches by 18 inches.

Ex. 351. A square and a rectangle have the same perimeter, 100 yards. The length of the rectangle is 4 times its breadth. Compare their areas.

Ex. 352. On a certain map the linear scale is 1 inch to 5 miles. How many acres are represented on this map by a square the perimeter of which is 1 inch?

PROPOSITION III. THEOREM.

398. *The area of a rectangle is equal to the product of its base by its altitude.*

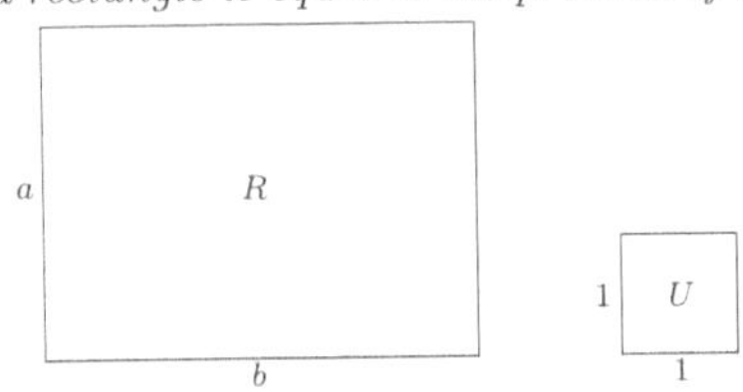

Let R be a rectangle, b its base, and a its altitude.

To prove that

$$\text{the area of } R = a \times b.$$

Proof. Let U be the unit of surface.

$$\frac{R}{U} = \frac{a \times b}{1 \times 1} = a \times b,$$

(*two rectangles are to each other as the products of their bases and altitudes*).

But

$$\frac{R}{U} = \text{the } \textit{number} \text{ of units of surface in } R. \qquad \S\ 393$$

$$\therefore \text{ the area of } R = a \times b. \qquad \text{Q.E.D.}$$

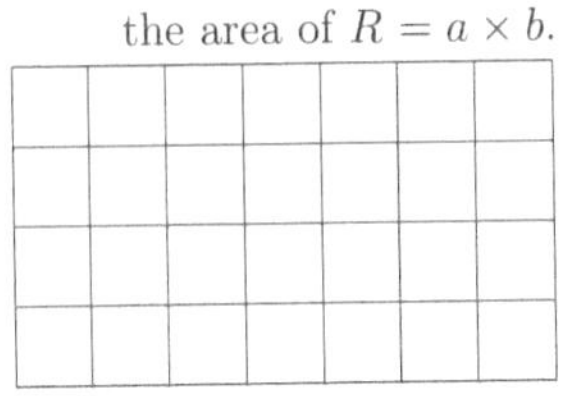

399. SCHOLIUM. When the base and altitude each contain the linear unit an integral number of times, this proposition is rendered evident by dividing the figure into squares, each equal to the unit of surface. Thus, if the base contains seven linear units, and the altitude four, the figure may be divided into twenty-eight squares, each equal to the unit of surface.

PROPOSITION IV. THEOREM.

400. *The area of a parallelogram is equal to the product of its base by its altitude.*

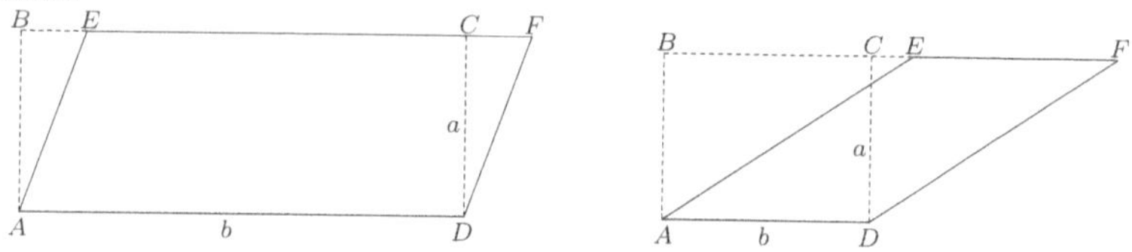

Let $AEFD$ be a parallelogram, b its base, and a its altitude.

To prove that the area of the ▱$AEFD = a \times b$.

Proof. From A draw $AB \parallel$ to DC to meet FE produced.

Then the figure $ABCD$ is a rectangle, with the same base and the same altitude as the ▱$AEFD$.

The rt. $_sABE$ and DCF are equal. § 151

For $AB = CD$, and $AE = DF$. § 178

From $ABFD$ take the DCF; the rect. $ABCD$ is left.

From $ABFD$ take the ABE; the ▱$AEFD$ is left.

rect. $ABCD$ ▱$AEFD$ Ax. 3

But the area of the rect. $ABCD = a \times b$. § 398

the area of the ▱$AEFD = a \times b$. Ax. 1

Q.E.D.

401. COR. 1. *Parallelograms having equal bases and equal altitudes are equivalent.*

402. COR. 2. *Parallelograms having equal bases are to each other as their altitudes; parallelograms having equal altitudes are to each other as their bases; any two parallelograms are to each other as the products of their bases by their altitudes.*

PROPOSITION V. THEOREM.

403. *The area of a triangle is equal to half the product of its base by its altitude.*

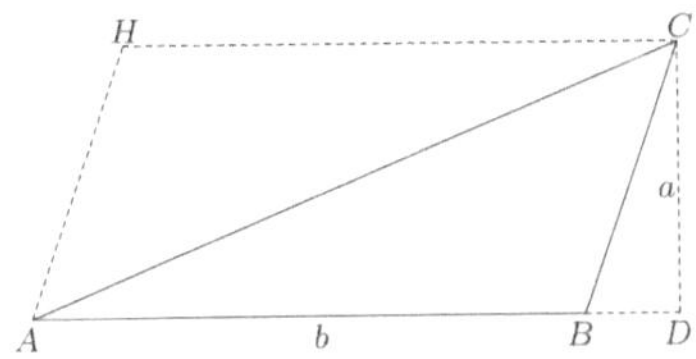

Let a be the altitude and b the base of the triangle ABC.

To prove that the area of the $ABC = \frac{1}{2}a \times b$.

Proof. Construct on AB and BC the parallelogram $ABCH$.

Then

$ABC = \frac{1}{2}$▱$ABCH$. § 179

The area of the ▱$ABCH = a \times b$. § 400

Therefore, the area of $ABC = \frac{1}{2}a \times b$. Ax. 7

Q.E.D.

404. COR. 1. *Triangles having equal bases and equal altitudes are equivalent.*

405. COR. 2. *Triangles having equal bases are to each other as their altitudes; triangles having equal altitudes are to each other as their bases; any two triangles are to each other as the products of their bases by their altitudes.*

406. COR. 3. *The product of the legs of a right triangle is equal to the product of the hypotenuse by the altitude from the vertex of the right angle.*

Ex. 353. The lines which join the middle point of either diagonal of a quadrilateral to the opposite vertices divide the quadrilateral into two equivalent parts.

PROPOSITION VI. THEOREM.

407. *The area of a trapezoid is equal to half the sum of its bases multiplied by the altitude.*

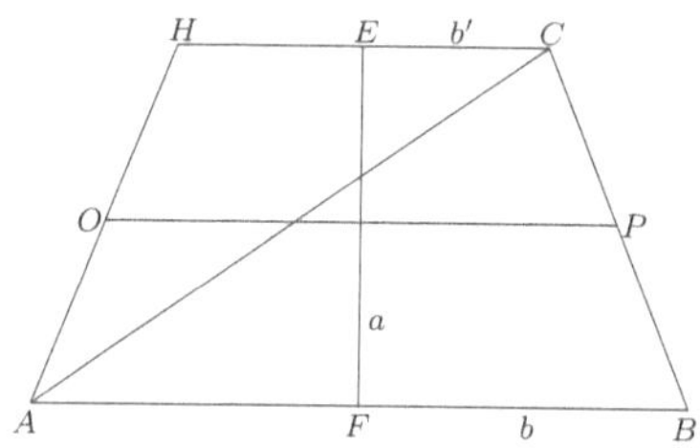

Let b and b' be the bases and a the altitude of the trapezoid $ABCH$.

To prove that the area of the $ABCH = \frac{1}{2}a(b + b')$.

Proof.

Draw the diagonal AC.

Then

the area of the $ABC = \frac{1}{2}a \times b$,

and

the area of the $AHC = \frac{1}{2}a \times b'$. § 403

the area of $ABCH = \frac{1}{2}a(b + b')$. Ax. 2

Q.E.D.

408. COR. *The area of a trapezoid is equal to the product of the median by the altitude.* § 190

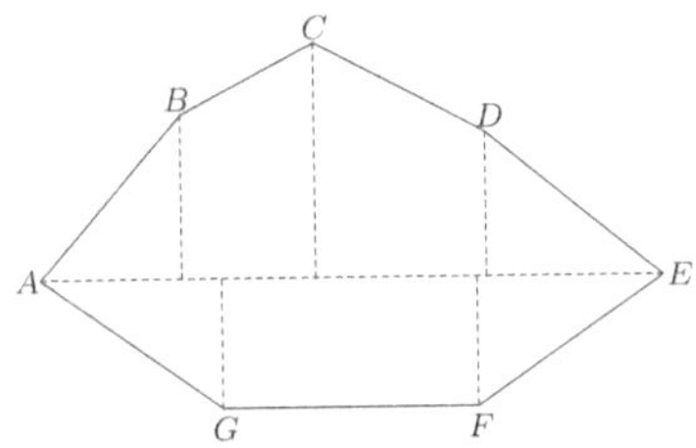

409. SCHOLIUM. The area of an irregular polygon may be found by dividing the polygon into triangles, and by finding the area of each of these triangles separately. Or, we may draw the longest diagonal, and let fall perpendiculars upon this diagonal from the other vertices of the polygon.

The sum of the areas of the right triangles, rectangles, and trapezoids thus formed is the area of the polygon.

PROPOSITION VII. THEOREM.

410. *The areas of two triangles which have an angle of the one equal to an angle of the other are to each other as the products of the sides including the equal angles.*

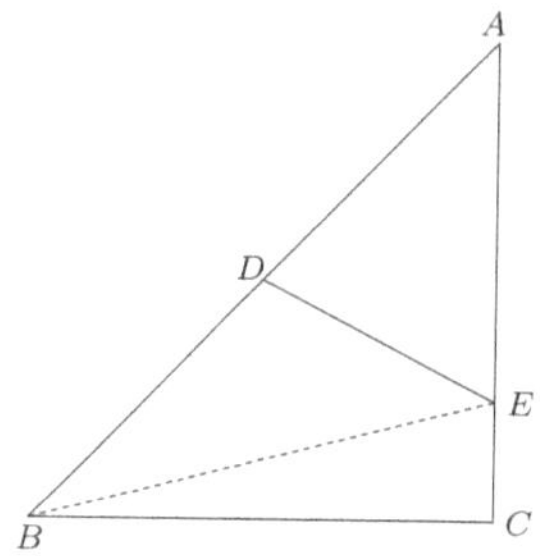

Let the triangles ABC and ADE have the common angle A.

To prove that $\dfrac{ABC}{ADE} = \dfrac{AB \times AC}{AD \times AE}$.

Proof.

Draw BE.

Now

$$\frac{ABC}{ABE} = \frac{AC}{AE},$$

and

$$\frac{ABE}{ADE} = \frac{AB}{AD}. \qquad \S\ 405$$

The products of the first members and of the second members of these equalities give

$$\frac{ABC}{ADE} = \frac{AB \times AC}{AD \times AE}. \qquad \text{Q.E.D.}$$

Ex. 354. The areas of two triangles which have an angle of the one supplementary to an angle of the other are to each other as the products of the sides including the supplementary angles.

COMPARISON OF POLYGONS.

PROPOSITION VIII. THEOREM.

411. *The areas of two similar triangles are to each as the squares of any two homologous sides.*

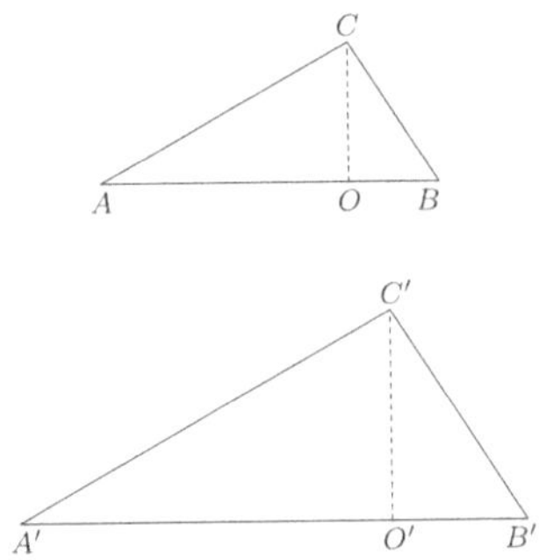

Let the two similar triangles be ACB and $A'C'B'$.

To prove that

$$\frac{ACB}{A'C'B'} = \frac{\overline{AB}^2}{\overline{A'B'}^2}.$$

Proof. Draw the altitudes CO and $C'O'$.

Then

$$\frac{ACB}{A'C'B'} = \frac{AB \times CO}{A'B' \times C'O'} = \frac{AB}{A'B'} \times \frac{CO}{C'O'}, \qquad \S\ 405$$

(*two* $_s$ *are to each other as the products of their bases by their altitudes*).

But

$$\frac{AB}{A'B'} = \frac{CO}{C'O'}. \qquad \S\ 361$$

(*the homologous altitudes of two similar* $_s$ *have the same ratio as any two homologous sides*).

Substitute, in the above equality, for $\frac{CO}{C'O'}$ its equal $\frac{AB}{A'B'}$;

then

$$\frac{ACB}{A'C'B'} = \frac{AB}{A'B'} \times \frac{AB}{A'B'} = \frac{\overline{AB}^2}{\overline{A'B'}^2}. \qquad \text{Q.E.D.}$$

Ex. 355. Prove this proposition by § 410.

Proposition IX. Theorem.

412. *The areas of two similar polygons are to each other as the squares of any two homologous sides.*

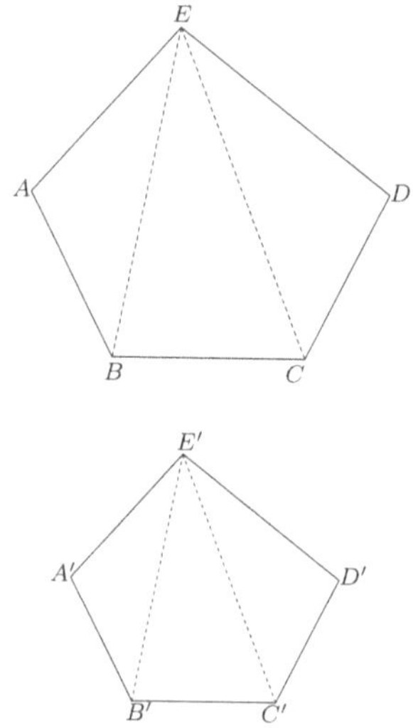

Let S and S' denote the areas of the two similar polygons ABC etc. and $A'B'C'$ etc.

To prove that

$$S : S' = \overline{AB}^2 : \overline{A'B'^2}.$$

Proof. By drawing all the diagonals from any homologous vertices E and E', the two similar polygons are divided into similar triangles. § 365

$$\frac{\overline{AB}^2}{\overline{A'B'^2}} = \frac{ABE}{A'B'E'} = \left(\frac{\overline{BE}^2}{\overline{B'E'^2}}\right) = \frac{BCE}{B'C'E'} = \text{etc.} \qquad \S\ 411$$

That is,

$$\frac{ABE}{A'B'E'} = \frac{BCE}{B'C'E'} = \frac{CDE}{C'D'E'}.$$

$$\frac{ABE + \ BCE + \ CDE}{A'B'E' + \ B'C'E' + \ C'D'E'} = \frac{ABE}{A'B'E'} = \frac{\overline{AB}^2}{\overline{A'B'^2}}. \qquad \S\ 335$$

$$S : S' = \overline{AB}^2 : \overline{A'B'^2}$$

Q.E.D.

413. Cor. 1. *The areas of two similar polygons are to each other as the squares of any two homologous lines.*

414. Cor. 2. *The homologous sides of two similar polygons have the same ratio as the square roots of their areas.*

Proposition X. Theorem.

415. *The square on the hypotenuse of a right triangle is equivalent to the sum of the squares on the two legs.*

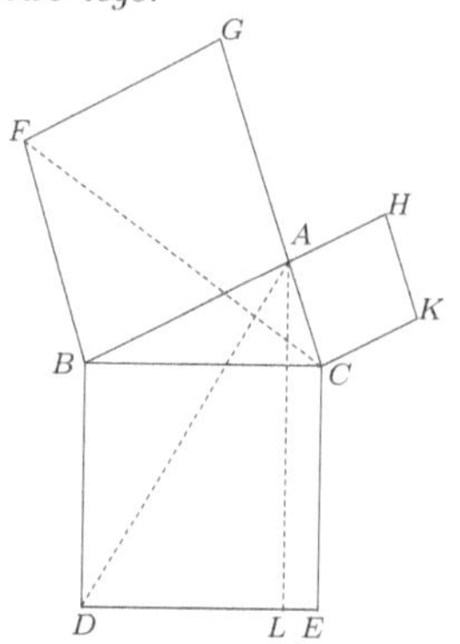

Let BE, CH, AF, be squares on the three sides of the right triangle ABC.

To prove that $BE \quad CH + AF$.

Proof. Through A draw $AL \parallel$ to CE, and draw AD and CF.

Since $_sBAC$, BAG, and CAH are rt. $_s$, CAG and BAH are straight lines. § 90

The

$$ABD = \quad FBC. \qquad § 143$$

For

$$BD = BC,$$

$$BA = BF, \qquad § 168$$

and

$$ABD = \quad FBC, \qquad \text{Ax. 2}$$

(*each being the sum of a rt.* *and the* ABC).

Now the rectangle BL is double the ABD,
(*having the same base* BD, *and the same altitude, the distance between the* $\parallel_s$ AL *and* BD),

and the square AF is double the FBC,
(*having the same base* FB, *and the same altitude* AB).

the rectangle BL is equivalent to the square AF. Ax. 6

In like manner, by drawing AE and BK, it may be proved that the rectangle CL is equivalent to the square CH.

Hence, the square BE, the sum of the rectangles BL and CL, is equivalent to the sum of the squares CH and AF. Q.E.D.

416. Cor. *The square on either leg of a right triangle is equivalent to the difference of the square on the hypotenuse and the square on the other leg.*

THEOREMS.

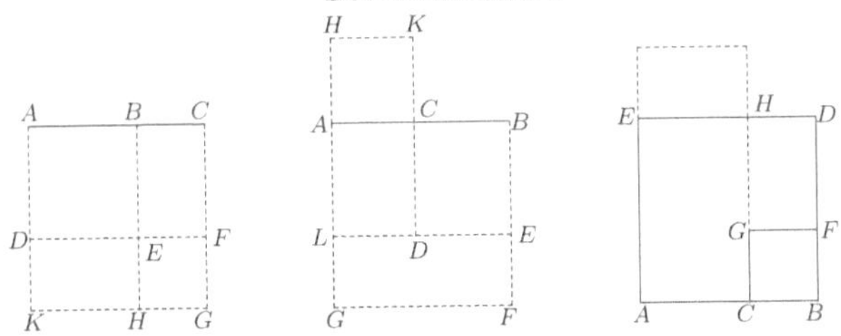

Ex. 356. The square constructed upon the sum of two straight lines is equivalent to the sum of the squares constructed upon these two lines, increased by twice the rectangle of these lines:

Let AB and BC be the two straight lines, and AC their sum. Construct the squares $ACGK$ and $ABED$ upon AC and AB, respectively. Prolong BE and DE until they meet KG and CG, respectively. Then we have the square $EFGH$, with sides each equal to BC. Hence, the square $ACGK$ is the sum of the squares $ABED$ and $EFGH$, and the rectangles $DEHK$ and $BCFE$, the dimensions of which are equal to AB and BC.

Ex. 357. The square constructed upon the difference of two straight lines is equivalent to the sum of the squares constructed upon these two lines, diminished by twice the rectangle of these lines.

Let AB and AC be the two straight lines, and BC their difference. Construct the square $ABFG$ upon AB, the square $ACKH$ upon AC, and the square $BEDC$ upon BC (as shown in the figure). Prolong ED to meet AG in L.

The dimensions of the rectangles $LEFG$ and $HKDL$ are AB and AC, and the square $BCDE$ is evidently the difference between the whole figure and the sum of these rectangles; that is, the square constructed upon BC is equivalent to the sum of the squares constructed upon AB and AC, diminished by twice the rectangle of AB and AC.

Ex. 358. The difference between the squares constructed upon two straight lines is equivalent to the rectangle of the sum and difference of these lines.

Let $ABDE$ and $BCFG$ be the squares constructed upon the two straight lines AB and BC. The difference between these squares is the polygon $ACGFDE$, which is composed of the rectangles $ACHE$ and $GFDH$. Prolong AE and CH to I and K, respectively, making EI and HK each equal to BC, and draw IK. The rectangles $GFDH$ and $EHKI$ are equal. The difference between the squares $ABDE$ and $BCGF$ is then equivalent to the rectangle $ACKI$, which has for dimensions AI, equal to $AB + BC$, and EH, equal to $AB - BC$.

Ex. 359. The area of a rhombus is equal to half the product of its diagonals.

Ex. 360. Two isosceles triangles are equivalent if their legs are equal each to each, and the altitude of one is equal to half the base of the other.

Ex. 361. The area of a circumscribed polygon is equal to half the product of its perimeter by the radius of the inscribed circle.

Ex. 362. Two parallelograms are equal if two adjacent sides of the one are equal, respectively, to two adjacent sides of the other, and the included angles are supplementary.

Ex. 363. If ABC is a right triangle, C the vertex of the right angle, BD a line cutting AC in D, then $\overline{BD}^2 + \overline{AC}^2 = \overline{AB}^2 + \overline{DC}^2$.

Ex. 364. Upon the sides of a right triangle as homologous sides three similar polygons are constructed. Prove that the polygon upon the hypotenuse is equivalent to the sum of the polygons upon the legs.

Ex. 365. If the middle points of two adjacent sides of a parallelogram are joined, a triangle is formed which is equivalent to one eighth of the parallelogram.

Ex. 366. If any point within a parallelogram is joined to the four vertices, the sum of either pair of triangles having parallel bases is equivalent to half the parallelogram.

Ex. 367. Every straight line drawn through the intersection of the diagonals of a parallelogram divides the parallelogram into two equal parts.

Ex. 368. The line which joins the middle points of the bases of a trapezoid divides the trapezoid into two equivalent parts.

Ex. 369. Every straight line drawn through the middle point of the median of a trapezoid cutting both bases divides the trapezoid into two equivalent parts.

Ex. 370. If two straight lines are drawn from the middle point of either leg of a trapezoid to the opposite vertices, the triangle thus formed is equivalent to half the trapezoid.

Ex. 371. The area of a trapezoid is equal to the product of one of the legs by the distance from this leg to the middle point of the other leg.

Ex. 372. The figure whose vertices are the middle points of the sides of any quadrilateral is equivalent to half the quadrilateral.

PROBLEMS OF CONSTRUCTION.

PROPOSITION XI. PROBLEM.

417. *To construct a square equivalent to the sum of two given squares.*

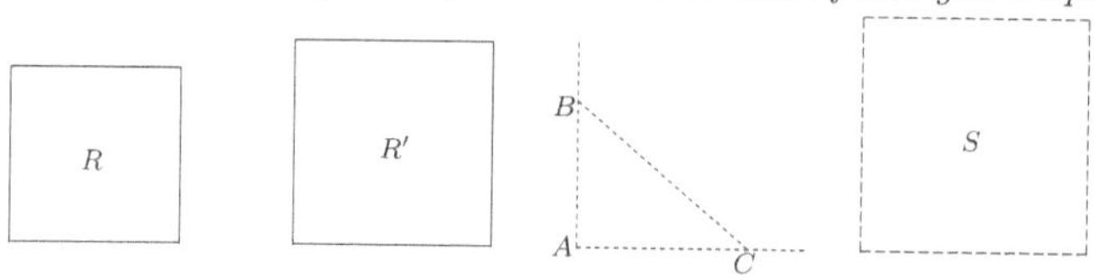

Let R and R' be two given squares.

To construct a square equivalent to $R' + R$.

Construct the rt. A.

Take AC equal to a side of R',

and AB equal to a side of R; and draw BC.

Construct the square S, having each of its sides equal to BC.

Then

S is the square required.

Proof.

$$\overline{BC}^2 \quad \overline{AC}^2 + \overline{AB}^2, \qquad \S\ 415$$

(*the square on the hypotenuse of a rt. is equivalent to the sum of the squares on the two legs*).

$$S \quad R' + R.$$

Q.E.F.

Ex. 373. If the perimeter of a rectangle is 72 feet, and the length is equal to twice the width, find the area.

Ex. 374. How many tiles 9 inches long and 4 inches wide will be required to pave a path 8 feet wide surrounding a rectangular court 120 feet long and 36 feet wide?

Ex. 375. The bases of a trapezoid are 16 feet and 10 feet; each leg is equal to 5 feet. Find the area of the trapezoid.

Proposition XII. Problem.

418. *To construct a square equivalent to the difference of two given squares.*

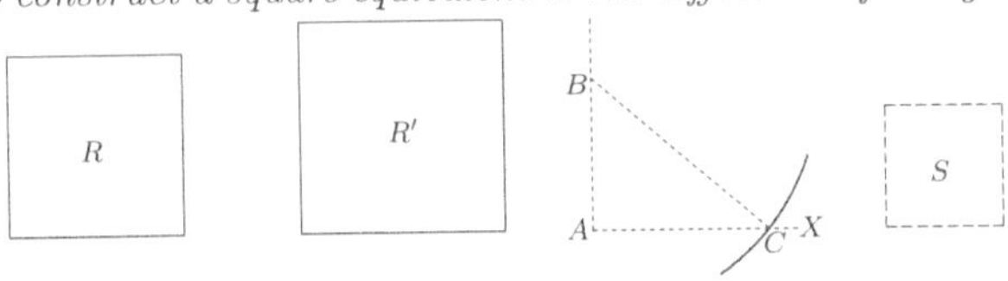

Let R be the smaller square and R' the larger.

To construct a square equivalent to $R' - R$.

Construct the rt. A.

Take AB equal to a side of R.

From B as a centre, with a radius equal to a side of R',

describe an arc cutting the line AX at C.

Construct the square S, having each of its sides equal to AC.

Then

S is the square required.

Proof.

$$\overline{AC}^2 \quad \overline{BC}^2 - \overline{AB}^2, \qquad \S\ 416$$

(*the square on either leg of a rt.* *is equivalent to the difference of the square on the hypotenuse and the square on the other leg*).

$$S \quad R' - R.$$

Q.E.F.

Ex. 376. Construct a square equivalent to the sum of two squares whose sides are 3 inches and 4 inches.

Ex. 377. Construct a square equivalent to the difference of two squares whose sides are $2\frac{1}{2}$ inches and 2 inches.

Ex. 378. Find the side of a square equivalent to the sum of two squares whose sides are 24 feet and 32 feet.

Ex. 379. Find the side of a square equivalent to the difference of two squares whose sides are 24 feet and 40 feet.

Ex. 380. A rhombus contains 100 square feet, and the length of one diagonal is 10 feet. Find the length of the other diagonal.

PROPOSITION XIII. PROBLEM.

419. *To construct a polygon similar to two given similar polygons and equivalent to their sum.*

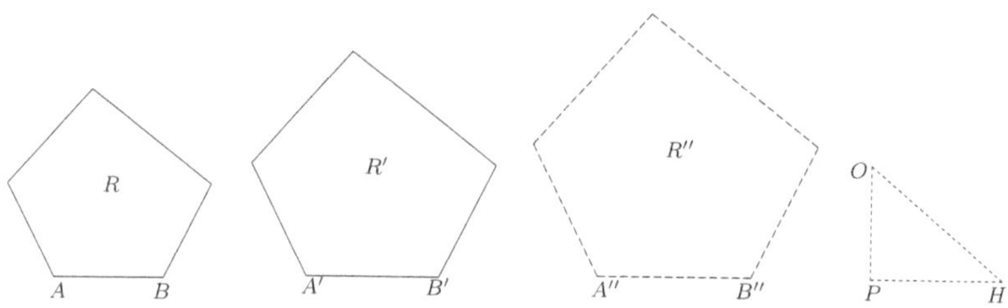

Let R and R' be two similar polygons, and AB and $A'B'$ two homologous sides.

To construct a similar polygon equivalent to $R+R'$.

Construct the rt. P.

Take PH equal to $A'B'$, and PO equal to AB.

Draw OH, and take $A''B''$ equal to OH.

Upon $A''B''$, homologous to AB, construct R'' similar to R.

Then R'' is the polygon required.

Proof.

$$\overline{PO}^2 + \overline{PH}^2 = \overline{OH}^2. \qquad \S\ 415$$

Put for PO, PH, and OH their equals AB, $A'B'$, and $A''B''$.
Then

$$\overline{AB}^2 + \overline{A'B'^2} = \overline{A''B''^2}.$$

Now

$$\frac{R}{R''} = \frac{\overline{AB}^2}{\overline{A''B''^2}}, \text{ and } \frac{R'}{R''} = \frac{\overline{A'B'^2}}{\overline{A''B''^2}}. \qquad \S\ 412$$

By addition,

$$\frac{R+R'}{R''} = \frac{\overline{AB}^2 + \overline{A'B'^2}}{\overline{A''B''^2}} = 1. \qquad \text{Ax. 2}$$

$$R'' \quad R+R'. \qquad \text{Q.E.F.}$$

Proposition XIV. Problem.

420. *To construct a triangle equivalent to a given polygon.*

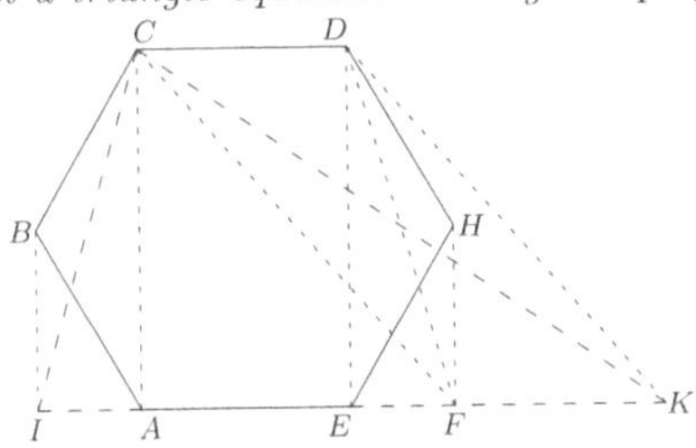

Let $ABCDHE$ be the given polygon.

To construct a triangle equivalent to the given polygon.

Let D, H, and E be any three consecutive vertices of the polygon. Draw the diagonal DE.

From H draw $HF \parallel$ to DE.

Produce AE to meet HF at F, and draw DF.

Again, draw CF, and draw $DK \parallel$ to CF to meet AF produced at K, and draw CK.

In like manner continue to reduce the number of sides of the polygon until we obtain the CIK.

Then CIK is the triangle required.

Proof. The polygon $ABCDF$ has one side less than the polygon $ACBDHE$, but the two polygons are equivalent.

For the part $ACBDE$ is common,

and the DEF DEH, § 404

(for the base DE is common, and their vertices F and H are in the line $FH \parallel$ to the base).

In like manner it may be proved that

$ABCK$ $ABCDF$, and CIK $ABCK$. Q.E.F.

Proposition XV. Problem.

421. *To construct a square equivalent to a given parallelogram.*

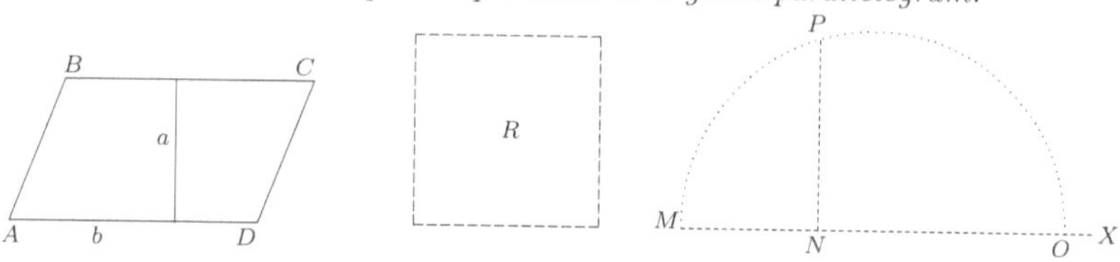

Let $ABCD$ be the parallelogram, b its base, and a its altitude.

To construct a square equivalent to the $\square ABCD$.

Upon a line MX take MN equal to a, NO equal to b.

Upon MO as a diameter, describe a semicircle.

At N erect NP to MO, meeting the circumference at P.

Then the square R, constructed upon a line equal to NP, is equivalent to the $\square ACBD$.

Proof.

$$MN : NP = NP : NO, \qquad \S\ 370$$

(*a let fall from any point of a circumference to the diameter is the mean proportional between the segments of the diameter*).

$$\overline{NP}^2 = MN \times NO = a \times b. \qquad \S\ 327$$

Therefore,

$$R \quad \square ABCD. \qquad \text{Q.E.F.}$$

422. Cor. 1. *A square may be constructed equivalent to a given triangle, by taking for its side the mean proportional between the base and half the altitude of the triangle.*

423. Cor. 2. *A square may be constructed equivalent to a given polygon, by first reducing the polygon to an equivalent triangle, and then constructing a square equivalent to the triangle.*

PROPOSITION XVI. PROBLEM.

424. *To construct a parallelogram equivalent to a given square, and having the sum of its base and altitude equal to a given line.*

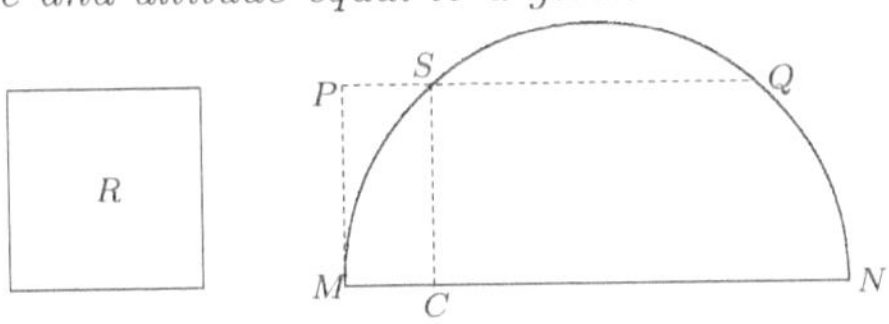

Let R be the given square, and let the sum of the base and altitude of the required parallelogram be equal to the given line MN.

To construct a ▱ equivalent to R, with the sum of its base and altitude equal to MN.

Upon MN as a diameter, describe a semicircle.

At M erect MP, a   to MN, equal to a side of the given square R.

Draw $PQ \parallel$ to MN, cutting the circumference at S.

Draw SC to MN.

Any ▱ having CM for its altitude and CN for its base is equivalent to R.

Proof.

$$SC = PM. \qquad \text{§§ 104, 180}$$

$$\overline{SC}^2 = \overline{PM}^2 = R.$$

$$MC : SC = SC : CN, \qquad \text{§ 370}$$

(*a   let fall from any point of a circumference to the diameter is the mean proportional between the segments of the diameter*).

Then

$$\overline{SC}^2 \quad MC \times CN. \qquad \text{§ 327}$$

Q.E.F.

NOTE. This problem may be stated as follows:

To construct two straight lines the sum and product of which are known.

Proposition XVII. Problem.

425. *To construct a parallelogram equivalent to a given square, and having the difference of its base and altitude equal to a given line.*

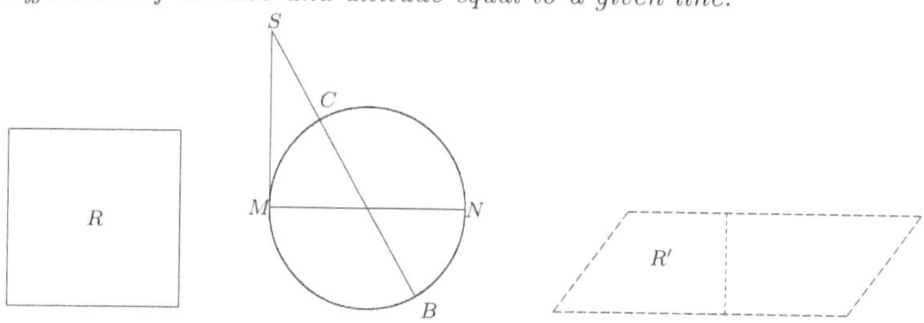

Let R be the given square, and let the difference of the base and altitude of the required parallelogram be equal to the given line MN.

To construct a ▱ equivalent to R, with the difference of its base and altitude equal to MN.

Upon the given line MN as a diameter, describe a circle.

From M draw MS, tangent to the ⊙, and equal to a side of the given square R.

Through the centre of the ⊙ draw SB intersecting the circumference at C and B.

Then any ▱, as R', having SB for its base and SC for its altitude, is equivalent to R.

Proof.

$$SB : SM = SM : SC, \qquad \text{§ 381}$$

(*if from a point without a ⊙ a secant and a tangent are drawn, the tangent is the mean proportional between the whole secant and the external segment*).

Then

$$\overline{SM}^2 \quad SB \times SC, \qquad \text{§ 327}$$

and the difference between SB and SC is the diameter of the ⊙, that is, MN.

Q.E.F.

Note. This problem may be stated: *To construct two straight lines the difference and product of which are known.*

PROPOSITION XVIII. PROBLEM.

426. *To construct a polygon similar to a given polygon P and equivalent to a given polygon Q.*

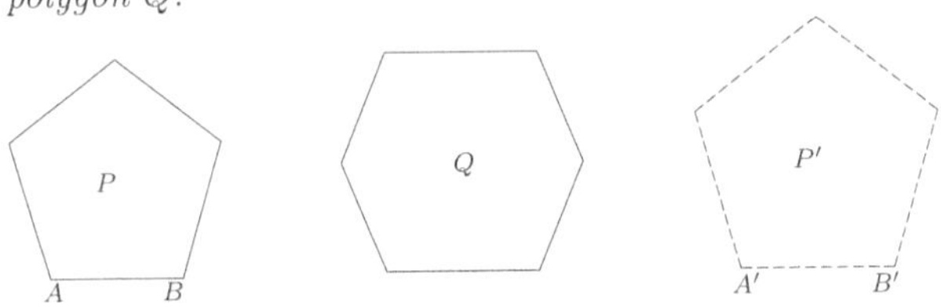

Let P and Q be the two given polygons, and AB a side of P.

To construct a polygon similar to P and equivalent to Q.

Find squares equivalent to P and Q, § 423

and let m and n respectively denote their sides.

Find $A'B'$, the fourth proportional to m, n, and AB. § 386

Upon $A'B'$, homologous to AB, construct P' similar to P.

Then

$$P' \quad Q.$$

Proof.

$$m : n = AB : A'B'. \qquad \text{Const.}$$

$$m^2 : n^2 = \overline{AB}^2 : \overline{A'B'}^2. \qquad \text{§ 338}$$

But

$$P \quad m^2, \text{ and } Q \quad n^2. \qquad \text{Const.}$$

$$P : Q = m^2 : n^2 = \overline{AB}^2 : \overline{A'B'}^2.$$

But

$$P : P' = \overline{AB}^2 : \overline{A'B'}^2. \qquad \text{§ 412}$$

$$P : Q = P : P'. \qquad \text{Ax. 1}$$

$$P' \quad Q. \qquad \text{Q.E.F.}$$

Ex. 381. To construct a square equivalent to the sum of any number of given squares.

Ex. 382. To construct a polygon similar to two given similar polygons and equivalent to their difference.

Proposition XIX. Problem.

427. *To construct a square which shall have a given ratio to a given square.*

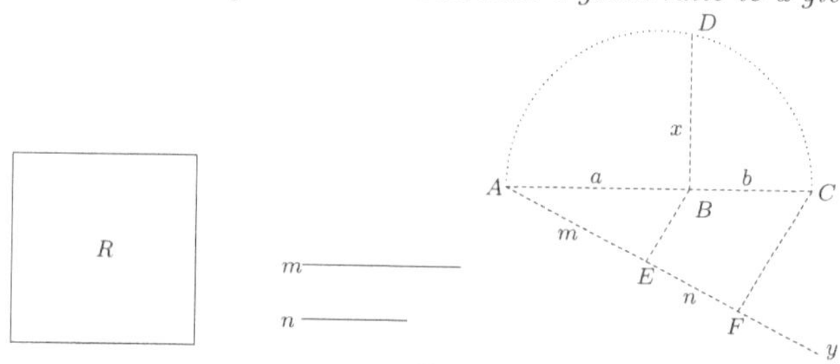

Let R be the given square, and $\frac{n}{m}$ the given ratio.

To construct a square which shall be to R as n is to m.

Take AB equal to a side of R, and draw Ay, making any convenient angle with AB.

On Ay take AE equal to m, EF equal to n, and draw EB.

Draw $FC \parallel$ to EB meeting AB produced at C.

On AC as a diameter, describe a semicircle.

At B erect the BD, meeting the semicircumference at D.

Then BD is a side of the square required.

Proof.

Denote AB by a, BC by b, and BD by x.

Now

$$a : x = x : b. \qquad \text{§ 370}$$

Therefore,

$$a^2 : x^2 = a : b. \qquad \text{§ 337}$$

But

$$a : b = m : n. \qquad \text{§ 342}$$

Therefore,

$$a^2 : x^2 = m : n. \qquad \text{Ax. 1}$$

By inversion,

$$x^2 : a^2 = n : m. \qquad \text{§ 331}$$

Hence, the square on BD will have the same ratio to R as n has to m.

Q.E.F.

PROPOSITION XX. PROBLEM.

428. *To construct a polygon similar to a given polygon and having a given ratio to it.*

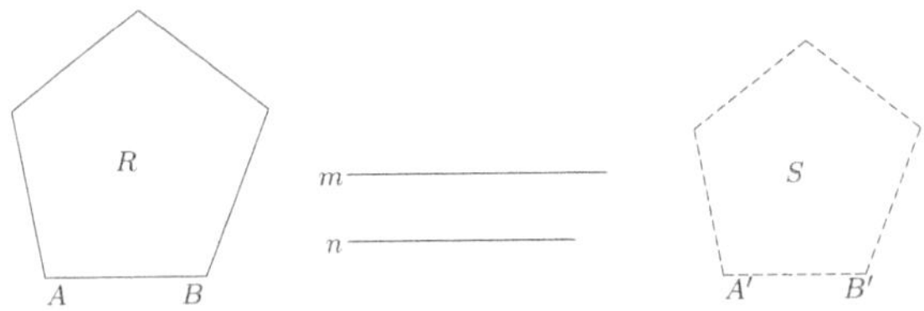

Let R be the given polygon, and $\frac{n}{m}$ the given ratio.

To construct a polygon similar to R, which shall be to R as n is to m.

Construct a line $A'B'$, such that the square on $A'B'$ shall be to the square on AB as n is to m. § 427

Upon $A'B'$, as a side homologous to AB, construct the polygon S similar to R. § 391

Then S is the polygon required.

Proof.

$$S : R = \overline{A'B'}^2 : \overline{AB}^2. \quad \text{§ 412}$$

But

$$\overline{A'B'}^2 : \overline{AB}^2 = n : m. \quad \text{Const.}$$

Therefore,

$$S : R = n : m. \quad \text{Ax. 1}$$

Q.E.F.

Ex. 383. To construct a triangle equivalent to a given triangle, and having one side equal to a given length l.

Ex. 384. To transform a triangle into an equivalent right triangle.

Ex. 385. To transform a given triangle into an equivalent right triangle, having one leg equal to a given length.

Ex. 386. To transform a given triangle into an equivalent right triangle, having the hypotenuse equal to a given length.

PROBLEMS OF CONSTRUCTION.

Ex. 387. To transform a triangle ABC into an equivalent triangle, having a side equal to a given length l, and an angle equal to angle BAC.

Upon AB (produced if necessary), take AD equal to l, draw $BE \parallel$ to CD, meeting AC (produced if necessary) at E.

BED $\quad$ BEC.

Ex. 388. To transform a given triangle into an equivalent isosceles triangle, having the base equal to a given length.

To construct a triangle equivalent to:

Ex. 389. The sum of two given triangles.

Ex. 390. The difference of two given triangles.

Ex. 391. To transform a given triangle into an equivalent equilateral triangle.

To transform a parallelogram into an equivalent:

Ex. 392. Parallelogram having one side equal to a given length.

Ex. 393. Parallelogram having one angle equal to a given angle.

Ex. 394. Rectangle having a given altitude.

To transform a square into an equivalent:

Ex. 395. Equilateral triangle.

Ex. 396. Right triangle having one leg equal to a given length.

Ex. 397. Rectangle having one side equal to a given length.

To construct a square equivalent to:

Ex. 398. Five eighths of a given square.

Ex. 399. Three fifths of a given pentagon.

Ex. 400. To divide a given triangle into two equivalent parts by a line through a given point P in one of the sides.

Ex. 401. To find a point within a triangle, such that the lines joining this point to the vertices shall divide the triangle into three equivalent parts.

Ex. 402. To divide a given triangle into two equivalent parts by a line parallel to one of the sides.

Ex. 403. To divide a given triangle into two equivalent parts by a line perpendicular to one of the sides.

PROBLEMS OF COMPUTATION.

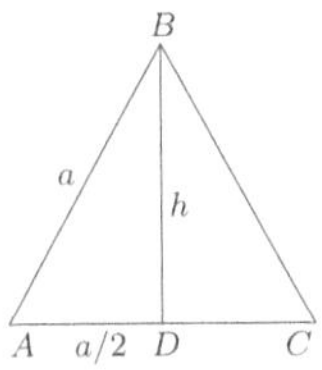

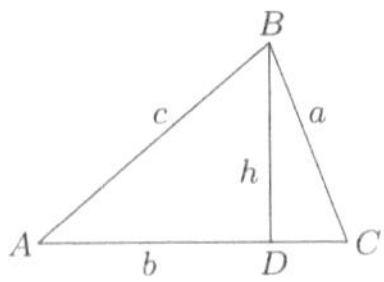

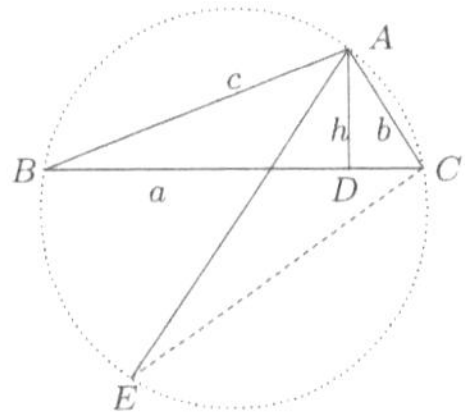

Ex. 404. To find the area of an equilateral triangle in terms of its side.

Denote the side by a, the altitude by h, and the area by S.

Then

$$h^2\ a^2 - \frac{a^2}{4} = \frac{3a^2}{4} = \frac{a^2}{4} \times 3. \qquad \S\ 372$$

$$h = \frac{a}{2}\sqrt{3}.$$

But

$$S = \frac{a \times h}{2}. \qquad \S\ 403$$

$$S = \frac{a}{2} \times \frac{a\sqrt{3}}{2} = \frac{a^2\sqrt{3}}{4}.$$

Ex. 405. To find the area of a triangle in terms of its sides.

By Ex. 312,

$$h = \frac{2}{b}\sqrt{s(s-a)(s-b)(s-c)}.$$

Hence,

$$S = \frac{b}{2} \times \frac{2}{b}\sqrt{s(s-a)(s-b)(s-c)} \qquad \S\ 403$$

$$= \sqrt{s(s-a)(s-b)(s-c)}.$$

Ex. 406. To find the area of a triangle in terms of the radius of the circumscribed circle.

If R denotes the radius of the circumscribed circle, and h the altitude of the triangle, we have, by § 384,

$$b \times c = 2R \times h.$$

Multiply by a, and we have,

$$a \times b \times c = 2R \times a \times h.$$

But

$$a \times h = 2S. \qquad \S\ 403$$

$$a \times b \times c = 4R \times S.$$

$$S = \frac{abc}{4R}.$$

Show that the radius of the circumscribed circle is equal to $\frac{abc}{4S}$.

Ex. 407. Find the area of a right triangle, if the length of the hypotenuse is 17 feet and the length of one leg is 8 feet.

Ex. 408. Find the ratio of the altitudes of two equivalent triangles, if the base of one is three times that of the other.

Ex. 409. The bases of a trapezoid are 8 feet and 10 feet, and the altitude is 6 feet. Find the base of the equivalent rectangle that has an equal altitude.

Ex. 410. Find the area of a rhombus, if the sum of its diagonals is 12 feet, and their ratio is 3 : 5.

Ex. 411. Find the area of an isosceles right triangle, if the hypotenuse is 20 feet.

Ex. 412. In a right triangle the hypotenuse is 13 feet, one leg is 5 feet. Find the area.

Ex. 413. Find the area of an isosceles triangle, if base $= b$, and leg $= c$.

Ex. 414. Find the area of an equilateral triangle, if one side $= 8$ feet.

Ex. 415. Find the area of an equilateral triangle, if the altitude $= h$.

Ex. 416. A house is 40 feet long, 30 feet wide, 25 feet high to the roof, and 35 feet high to the ridge-pole. Find the number of square feet in its entire exterior surface.

Ex. 417. The sides of a right triangle are as $3 : 4 : 5$. The altitude upon the hypotenuse is 12 feet. Find the area.

Ex. 418. Find the area of a right triangle, if one leg $= a$, and the altitude upon the hypotenuse $= h$.

Ex. 419. Find the area of a triangle, if the lengths of the sides are 104 feet, 111 feet, and 175 feet.

Ex. 420. The area of a trapezoid is 700 square feet. The bases are 30 feet and 40 feet, respectively. Find the altitude.

Ex. 421. $ABCD$ is a trapezium; $AB = 87$ feet, $BC = 119$ feet, $CD = 41$ feet, $DA = 169$ feet, $AC = 200$ feet. Find the area.

Ex. 422. What is the area of a quadrilateral circumscribed about a circle whose radius is 25 feet, if the perimeter of the quadrilateral is 400 feet? What is the area of a hexagon that has a perimeter of 400 feet and is circumscribed about the same circle of 25 feet radius (Ex. 361)?

Ex. 423. The base of a triangle is 15 feet, and its altitude is 8 feet. Find the perimeter of an equivalent rhombus, if the altitude is 6 feet.

Ex. 424. Upon the diagonal of a rectangle 24 feet by 10 feet a triangle equivalent to the rectangle is constructed. What is its altitude?

Ex. 425. Find the side of a square equivalent to a trapezoid whose bases are 56 feet and 44 feet, and each leg is 10 feet.

Ex. 426. Through a point P in the side AB of a triangle ABC, a line is drawn parallel to BC so as to divide the triangle into two equivalent parts. Find the value of AP in terms of AB.

Ex. 427. What part of a parallelogram is the triangle cut off by a line from one vertex to the middle point of one of the opposite sides?

Ex. 428. In two similar polygons, two homologous sides are 15 feet and 25 feet. The area of the first polygon is 450 square feet. Find the area of the second polygon.

Ex. 429. The base of a triangle is 32 feet, its altitude 20 feet. What is the area of the triangle cut off by a line parallel to the base at a distance of 15 feet from the base?

Ex. 430. The sides of two equilateral triangles are 3 feet and 4 feet. Find the side of an equilateral triangle equivalent to their sum.

Ex. 431. If the side of one equilateral triangle is equal to the altitude of another, what is the ratio of their areas?

Ex. 432. The sides of a triangle are 10 feet, 17 feet, and 21 feet. Find the areas of the parts into which the triangle is divided by the bisector of the angle formed by the first two sides.

Ex. 433. In a trapezoid, one base is 10 feet, the altitude is 4 feet, the area is 32 square feet. Find the length of a line drawn between the legs parallel to the bases and distant 1 foot from the lower base.

Ex. 434. The diagonals of a rhombus are 90 yards and 120 yards, respectively. Find the area, the length of one side, and the perpendicular distance between two parallel sides.

Ex. 435. Find the number of square feet of carpet that are required to cover a triangular floor whose sides are, respectively, 26 feet, 35 feet, and 51 feet.

Ex. 436. If the altitude h of a triangle is increased by a length m, how much must be taken from the base a that the area may remain the same?

Ex. 437. Find the area of a right triangle, having given the segments p, q, into which the hypotenuse is divided by a perpendicular drawn to the hypotenuse from the vertex of the right angle.

BOOK V. REGULAR POLYGONS AND CIRCLES.

429. Def. A **regular polygon** is a polygon which is both equilateral and equiangular. The equilateral triangle and the square are examples.

Proposition I. Theorem.

430. *An equilateral polygon inscribed in a circle is a regular polygon.*

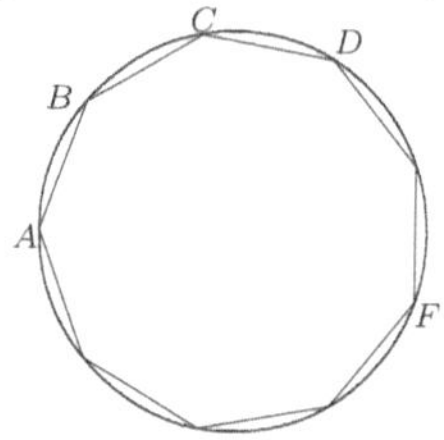

Let ABC etc. be an equilateral polygon inscribed in a circle.

To prove that the polygon ABC etc. is a regular polygon.

Proof.

The arcs AB, BC, CD, etc., are equal. § 243

Hence, arcs ABC, BCD, etc., are equal. Ax. 2

Therefore, arcs CFA, DFB, etc., are equal. Ax. 3

Therefore, $_sA$, B, C, etc., are equal. § 289

Therefore, the polygon ABC etc. is a regular polygon, being equilateral and equiangular. § 429

Q.E.D.

PROPOSITION II. THEOREM.

431. *A circle may be circumscribed about, and a circle may be inscribed in, any regular polygon.*

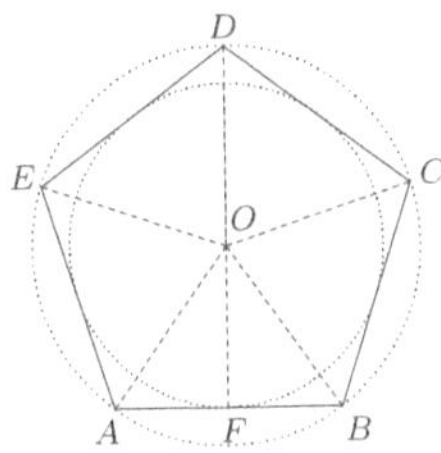

Let $ABCDE$ be a regular polygon.

1. *To prove that a circle may be circumscribed about $ABCDE$.*

Proof. Let O be the centre of the circle which may be passed through A, B, and C. § 258

Draw OA, OB, OC, and OD.

Then

$$ABC = BCD, \qquad \text{§ 429}$$

and

$$OBC = OCB. \qquad \text{§ 145}$$

By subtraction,

$$OBA = OCD. \qquad \text{Ax. 3}$$

The $_{s}OBA$ and OCD are equal. § 143

For

$$OBA = OCD,$$

$$OB = OC, \qquad \text{§ 217}$$

and

$$AB = CD. \qquad \text{§ 429}$$

$$OA = OD. \qquad \text{§ 128}$$

the circle passing through A, B, C, passes through D.

In like manner it may be proved that the circle passing through B, C, and D also passes through E; and so on.

Therefore, the circle described from O as a centre, with a radius OA, will be circumscribed about the polygon. § 231

2. *To prove that a circle may be inscribed in $ABCDE$.*

Proof. Since the sides of the regular polygon are equal chords of the circumscribed circle, they are equally distant from the centre. § 249

Therefore, the circle described from O as a centre, with the distance from O to a side of the polygon as a radius, will be inscribed in the polygon (§ 232). Q.E.D.

432. DEF. The radius of the circumscribed circle, OA, is called the **radius** of the polygon.

433. DEF. The radius of the inscribed circle, OF, is called the **apothem** of the polygon.

434. DEF. The common centre, O, of the circumscribed and inscribed circles is called the **centre** of the polygon.

435. DEF. The angle between radii drawn to the extremities of any side is called the **angle at the centre** of the polygon.

By joining the centre to the vertices of a regular polygon, the polygon can be decomposed into as many equal isosceles triangles as it has sides.

436. COR. 1. *The angle at the centre of a regular polygon is equal to four right angles divided by the number of sides of the polygon. Hence, the angles at the centre of any regular polygon are all equal.*

437. COR. 2. *The radius drawn to any vertex of a regular polygon bisects the angle at the vertex.*

438. COR. 3. *The angle at the centre of a regular polygon and an interior angle of the polygon are supplementary.*

For

$_sFOB$ and FBO are complementary. § 135

their doubles AOB and FBC are supplementary. Ax. 6

PROPOSITION III. THEOREM.

439. *If the circumference of a circle is divided into any number of equal arcs, the chords joining the successive points of division form a regular inscribed polygon; and the tangents drawn at the points of division form a regular circumscribed polygon.*

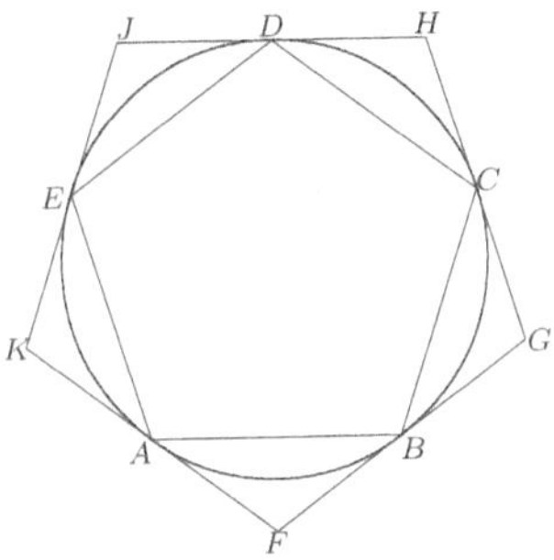

Suppose the circumference divided into equal arcs AB, BC, etc. Let AB, BC, etc., be the chords, FBG, GCH, etc., the tangents.

1. *To prove that $ABCDE$ is a regular polygon.*

Proof.

The sides AB, BC, CD, etc., are equal. § 241

Therefore, the polygon is regular. § 430

2. *To prove that To prove that $FGHIK$ is a regular polygon.*

Proof. The $_sAFB$, BGC, CHD, etc., are all equal isosceles triangles. §§ 295,139

$_sF$, G, H, etc., are equal, and FB, BG, GC, etc., are equal.

$FG = GH = HI$, etc. Ax. 6

$FGHIK$ is a regular polygon. § 429

Q.E.D.

440. COR. 1. *Tangents to a circle at the vertices of a regular inscribed polygon form a regular circumscribed polygon of the same number of sides as the inscribed polygon.*

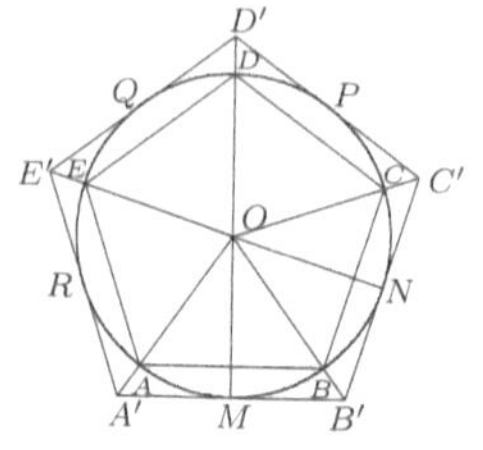

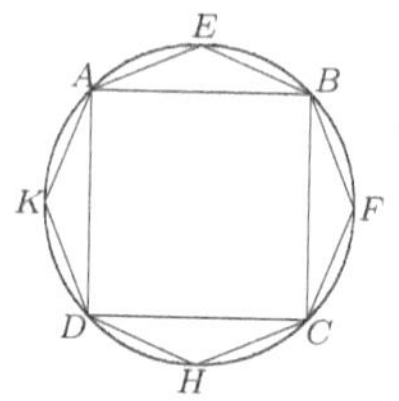

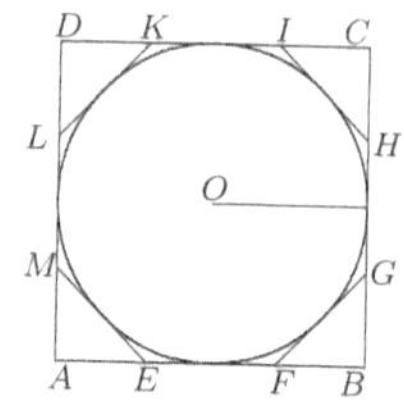

441. Cor. 2. *Tangents to a circle at the middle points of the arcs subtended by the sides of a regular inscribed polygon form a circumscribed regular polygon, whose sides are parallel to the sides of the inscribed polygon and whose vertices lie on the radii (prolonged) of the inscribed polygon.*

For two corresponding sides, AB and $A'B'$, are perpendicular to OM (§§ 248, 254), and are parallel (§ 104); and the tangents MB' and NB', intersecting at a point equidistant from OM and ON (§ 261), intersect upon the bisector of the MON (§ 162); that is, upon the radius OB.

442. Cor. 3. *If the vertices of a regular inscribed polygon are joined to the middle points of the arcs subtended by the sides of the polygon, the joining lines form a regular inscribed polygon of double the number of sides.*

443. Cor. 4. *Tangents at the middle points the arcs between adjacent points of contact of the sides of a regular circumscribed polygon form a regular circumscribed polygon of double the number of sides.*

444. Cor. 5. *The perimeter of an inscribed polygon is less than the perimeter of an inscribed polygon of double the number of sides; and the perimeter of a circumscribed polygon is greater than the perimeter of a circumscribed polygon of double the number of sides.*

For two sides of a triangle are together greater than the third side. § 138

PROPOSITION IV. THEOREM.

445. *Two regular polygons of the same number of sides are similar.*

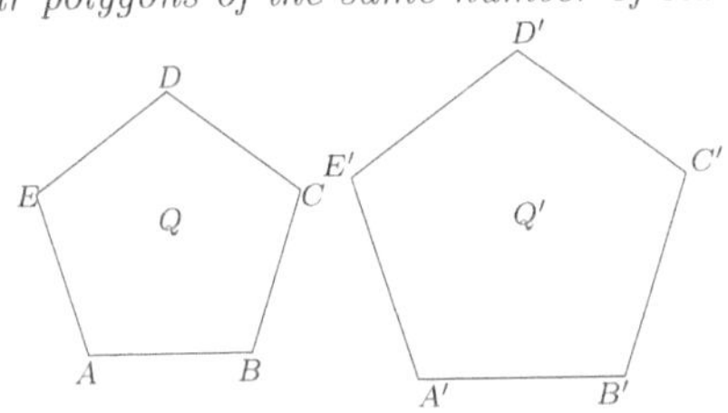

Let Q and Q' be two regular polygons, each having n sides.

To prove that Q and Q' are similar.

Proof. The sum of the interior s of each polygon is equal to

$$(n-2)2 \text{ rt. } s, \qquad \S\ 205$$

(*the sum of the interior s of a polygon is equal to 2 rt. s taken as many times less two as the polygon has sides*).

$$\text{Each angle of either polygon} = \frac{(n-2)2 \text{ rt. } s}{n}, \qquad \S\ 206$$

(*for the s of a regular polygon are all equal, and hence each is equal to the sum of the s divided by their number*).

Hence, the two polygons Q and Q' are mutually equiangular.

Since $AB = BC$, etc., and $A'B' = B'C'$, etc., § 429

$$AB : A'B' = BC : B'C', \text{ etc.}$$

Hence, the two polygons have their homologous sides proportional.

Therefore the two polygons are similar. § 351

Q.E.D.

446. COR. *The areas of two regular polygons of the same number of sides are to each other as the squares of any two homologous sides.* § 412

Proposition V. Theorem.

447. *The perimeters of two regular polygons of the same number of sides are to each other as the radii of their circumscribed circles, and also as the radii of their inscribed circles.*

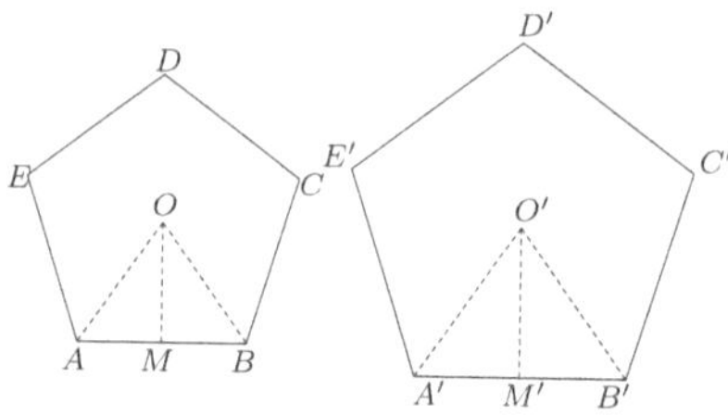

Let P and P' denote the perimeters, O and O' the centres, of the two regular polygons.

From O, O' draw OA, $O'A'$, OB, $O'B'$, and the ${}_s\, OM, O'M'$.

To prove that $P : P' = OA : O'A' = OM : O'M'$.

Proof.

Since the polygons are similar, § 445

$$P : P' = AB : A'B'.$$ § 364

The ${}_sOAB$ and $O'A'B'$ are isosceles. § 431

Now

$$O = O',$$ § 436

and

$$OA : OB = O'A' : O'B'.$$

the ${}_sOAB$ and $O'A'B'$ are similar. § 357

$$AB : A'B' = OA : O'A'.$$ § 351

Also,

$$AB : A'B' = OM : O'M'.$$ § 361

$$P : P' = OA : O'A' = OM : O'M'.$$ Ax. 1

Q.E.D.

448. Cor. *The areas of two regular polygons of the same number of sides are to each other as the squares of the radii of the circumscribed circles, and of the inscribed circles.* § 413

PROPOSITION VI. THEOREM.

449. *If the number of sides of a regular inscribed polygon is indefinitely increased, the apothem of the polygon approaches the radius of the circle as its limit.*

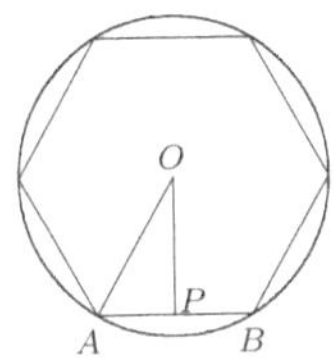

Let AB be a side and OP the apothem of a regular polygon of n sides inscribed in the circle whose radius is OA.

To prove that OP approaches OA as a limit, when n increases indefinitely.

Proof.

$$OP < OA, \quad \text{§ 97}$$

and

$$OA - OP < AP. \quad \text{§ 138}$$

$$OA - OP < AB, \text{ which is twice } AP. \quad \text{§ 245}$$

Now, if n is taken sufficiently great, AB, and consequently $OA - OP$, can be made less than any assigned value, however small, but cannot be made zero.

Since $OA - OP$ can be made less than any assigned value by increasing n, but cannot be made zero, OA is the limit of OP by the test for a limit. § 275

Q.E.D.

450. COR. *If the number of sides of a regular inscribed polygon is indefinitely increased, the square of the apothem approaches the square of the radius of the circle as a limit.*

For

$$\overline{OA}^2 - \overline{OP}^2 = \overline{AP}^2. \quad \text{§ 372}$$

But by taking n sufficiently great, AB and consequently AP, the half of AB, can be made less than any assigned value.

Therefore, $\overline{AP}^2$, the product of AP by AP, can be made less than any assigned value; for the product of two finite factors approaches zero as a limit, if *either* factor approaches zero as a limit (§ 276); and for a still stronger reason, the product approaches zero as a limit, if *each* of the factors approaches zero as a limit.

PROPOSITION VII. THEOREM.

451. *An arc of a circle is less than any line which envelops it and has the same extremities.*

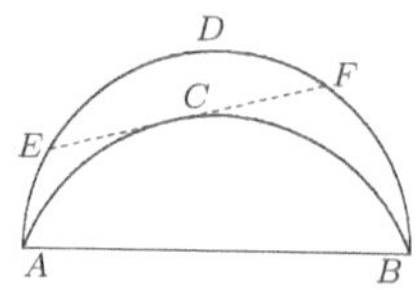

Let ACB be an arc of a circle, and AB its chord.

To prove that the arc ACB is less than any other line which envelops this arc and terminates at A and B.

Proof. Of all the lines that can be drawn, each to include the area ACB between itself and the chord AB, there must be at least one shortest line; for all the lines are not equal.

Now the enveloping line ADB cannot be the shortest; for drawing ECF tangent to the arc ACB at C, the line $AECFB < AEDFB$, since $ECF < EDF$. § 49

In like manner it can be shown that no other enveloping line can be the shortest. Therefore, ACB is the shortest.

Q.E.D.

452. COR. 1. *The circumference of a circle is less than the perimeter of any polygon circumscribed about it.*

453. COR. 2. *Any convex curve is less than the perimeter of a polygon circumscribed about it.*

PROPOSITION VIII. THEOREM.

454. *The circumference of a circle is the limit which the perimeters of regular inscribed polygons and of similar circumscribed polygons approach, if the number of sides of the polygons is indefinitely increased; and the area of a circle is the limit which the areas of these polygons approach.*

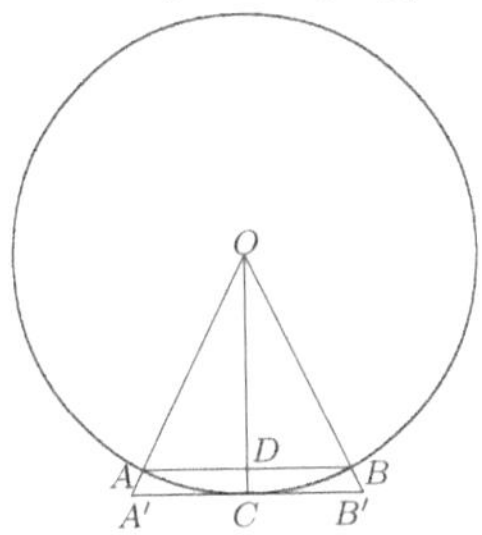

Let P and P' denote the lengths of the perimeters, AB and $A'B'$ two homologous sides, R and R' the radii, of the polygons, and C the circumference of the circle.

To prove that C is the limit of P and of P', if the number of sides of the polygons is indefinitely increased.

Proof.

Since the polygons are similar by hypothesis,

$$P' : P = R' : R. \qquad \text{§ 447}$$

Therefore,

$$P' - P : P = R' - R : R. \qquad \text{§ 333}$$

Whence,

$$R(P' - P) = P(R' - R). \qquad \text{§ 327}$$

Therefore,

$$P' - P = \tfrac{P}{R}(R' - R).$$

Now P is always less than C. § 273

$$P' - P < \tfrac{C}{R}(R' - R).$$

But $R' - R$, which is less than $A'C$ (§ 138), can be made less than any assigned quantity by increasing the number of sides of the polygons; and therefore $\frac{C}{R}(R' - R)$ can be made less than any assigned quantity. § 276

Hence, $P' - P$ can be made less than any assigned quantity.

Since P' is always greater than C (§ 452), and P is always less than C (§ 273), the difference between C and either P' or P is less than the difference $P' - P$, and consequently can be made less than any assigned quantity, but cannot be made zero.

Therefore, C is the common limit of P' and P. § 275

Let K denote the area of the circle, S the area of the inscribed polygon, and S' the area of the circumscribed polygon.

2. *To prove that K is the limit of S and S'.*

Proof.

$$S' : S = R'^2 : R^2.$$ § 448

By division,

$$S' - S : S = R'^2 - R^2 : R^2.$$ § 333

Whence

$$S' - S = \frac{S}{R^2}(R'^2 - R^2).$$

Now K is always greater than S. Ax. 8

Therefore,

$$S' - S < \frac{K}{R^2}(R'^2 - R^2).$$

But $R'^2 - R^2$, which is equal to $(R' + R)(R' - R)$, can be made less than any assigned quantity; and therefore $\frac{K}{R^2}(R'^2 - R^2)$ can be made less than any assigned quantity. § 276

Hence, $S' - S$ can be made less than any assigned quantity.

Since $S' > K$ always, and $S < K$ always (Ax. 8), the difference between K and either S' or S is less than the difference $S' - S$, and consequently can be made less than any assigned quantity, but cannot be made zero.

Therefore, K is the common limit of S' and S. § 275

Q.E.D.

PROPOSITION IX. THEOREM.

455. *Two circumferences have the same ratio as their radii.*

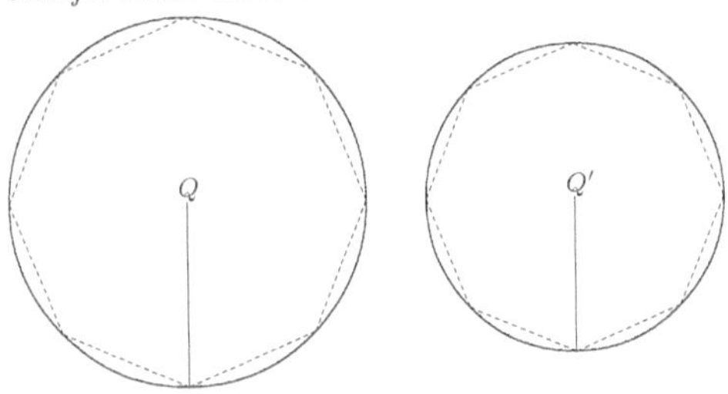

Let C and C' be the circumferences, R and R' the radii, of the two circles Q and Q'.

To prove that

$$C : C' = R : R'.$$

Proof. Inscribe in the $\odot_s$ two similar regular polygons, and denote their perimeters by P and P'.

Then

$$P : P' = R : R'. \qquad \S\ 447$$

Conceive the number of sides of these regular polygons to be indefinitely increased, the polygons continuing similar.

Then P and P' will have C and C' as limits. § 454

But $P : P'$ will always be equal to $R : R'$. § 447

$$C : C' = R : R'. \qquad \S\ 285$$

Q.E.D.

456. COR. *The ratio of the circumference of a circle to its diameter is constant.*

For

$$C : C' = R : R'. \qquad \S\ 455$$

$$C : C' = 2R : 2R'. \qquad \S\ 340$$

By alternation,

$$C : 2R = C' : 2R'. \qquad \S\ 330$$

457. DEF. The constant ratio of the circumference of a circle to its diameter is represented by the Greek letter π.

458. Cor. $\pi = \dfrac{C}{2R}$. $\quad C = 2\pi R$.

Proposition X. Theorem.

459. *The area of a regular polygon is equal to half the product of its apothem by its perimeter.*

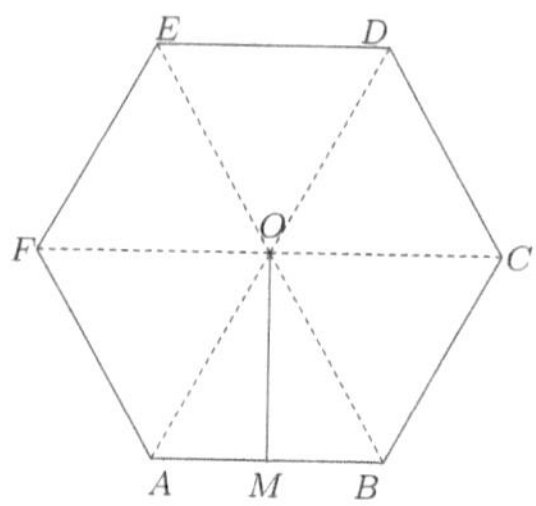

Let P represent the perimeter, R the apothem, and S the area of the regular polygon ABC etc.

To prove that $S = \frac{1}{2}R \times P$.

Proof.

Draw the radii OA, OB, OC, etc.

The polygon is divided into as many ${}_s$ as it has sides.

The apothem is the common altitude of these ${}_s$,

and the area of each $= \frac{1}{2}R$ multiplied by the base. § 403

Hence, the area of all the ${}_s$ is equal to $\frac{1}{2}R$ multiplied by the sum of all the bases.

But the sum of the areas of all the ${}_s$ is equal to the area of the polygon. Ax. 9

And the sum of all the bases of the ${}_s$ is equal to the perimeter of the polygon. Ax. 9

$$S = \tfrac{1}{2}R \times P.$$ Q.E.D.

460. Def. In different circles **similar arcs**, **similar sectors**, and **similar segments** are such as correspond to *equal angles at the centre.*

PROPOSITION XI. THEOREM.

461. *The area of a circle is equal to half the product of its radius by its circumference.*

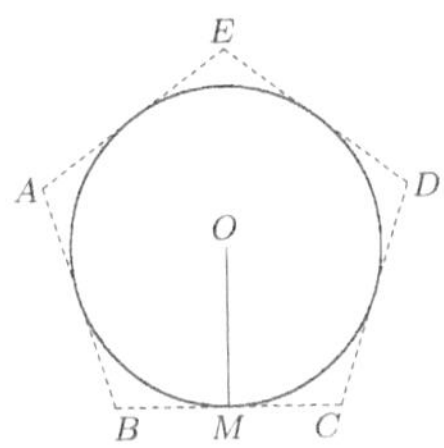

Let R represent the radius, C the circumference, and S the area, of the circle whose centre is O.

To prove that

$$S = \tfrac{1}{2}R \times C.$$

Proof. Circumscribe any regular polygon about the circle, and denote its perimeter by P, and its area by S'.

Then

$$S' = \tfrac{1}{2}R \times P. \qquad \S\ 459$$

Conceive the number of sides of the polygon to be indefinitely increased.

Then P approaches C as its limit, § 454

$\frac{1}{2}R \times P$ approaches $\frac{1}{2}R \times C$ as its limit, § 279

and S' approaches S as its limit. § 454

But

$$S' = \tfrac{1}{2}R \times P, \text{ always.} \qquad \S\ 459$$

$$S = \tfrac{1}{2}R \times C. \qquad \S\ 284$$

Q.E.D.

462. COR. 1. *The area of a sector is equal to half the product of its radius by its arc.*

For the sector and its arc are like parts of the circle and its circumference, respectively.

463. Cor. 2. *The area of a circle is equal to π times the square of its radius.*

For the area of the $\odot = \frac{1}{2}R \times C = \frac{1}{2}R \times 2\pi R = \pi R^2$.

464. Cor. 3. *The areas of two circles are to each other as the squares of their radii.*

For, if S and S' denote the areas, and R and R' the radii,

$$S : S' = \pi R^2 : \pi R'^2 = R^2 : R'^2.$$

465. Cor. 4. *Similar arcs are to each other as their radii; similar sectors are to each other as the squares of their radii.*

PROPOSITION XII. THEOREM.

466. *The areas of two similar segments are to each other as the squares of their radii.*

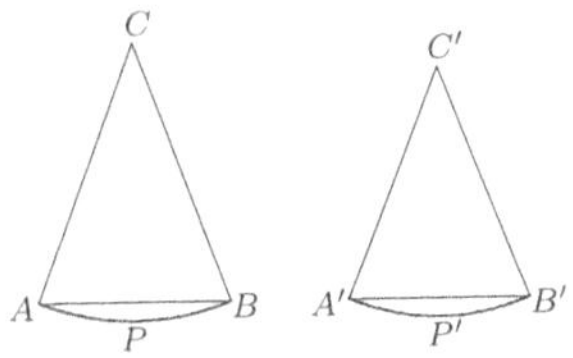

Let AC and $A'C'$ be the radii of the two similar sectors ACB and $A'C'B'$, and let ABP and $A'B'P'$ be the corresponding segments.

To prove that

$$ABP : A'B'P' = \overline{AC}^2 : \overline{A'C'}^2.$$

Proof.

Sector ACB : Sector $A'C'B' = \overline{AC}^2 : \overline{A'C'}^2$. § 465

The $\;_sACB$ and $A'C'B'$ are similar. § 357

$ACB : \quad A'C'B' = \overline{AC}^2 : \overline{A'C'}^2$. § 411

sector ACB : sector $A'C'B' = \quad ACB : \quad A'C'B'$. Ax. 1

sector $ACB : \quad ACB$ = sector $A'C'B' : \quad A'C'B'$. § 330

$$\frac{\text{sector } ACB - \quad ACB}{\text{sector } A'C'B' - \quad A'C'B'} = \frac{ACB}{A'C'B'} = \frac{\overline{AC}^2}{\overline{A'C'}^2}.$$ § 333

That is,

$$ABP : A'B'P' = \overline{AC}^2 : \overline{A'C'}^2.$$ Q.E.D.

PROBLEMS OF CONSTRUCTION.

Proposition XIII. Problem.

467. *To inscribe a square in a given circle.*

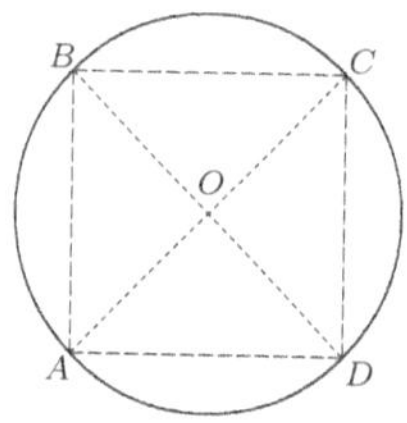

Let O be the centre of the given circle.

To inscribe a square in the given circle.

Draw two diameters AC and BD to each other.

Draw AB, BC, CD, and DA.

Then $ABCD$ is the square required.

Proof.

The ${}_s ABC$, BCD, etc., are rt. ${}_s$, § 290
(each being inscribed in a semicircle),

and the sides AB, BC, etc., are equal, § 241
(in the same ⊙ equal arcs are subtended by equal chords).

Hence the quadrilateral $ABCD$ is a square. § 168

Q.E.F.

468. Cor. *By bisecting the arcs AB, BC, etc., a regular polygon of eight sides may be inscribed in the circle; and, by continuing the process, regular polygons of sixteen, thirty-two, sixty-four, etc., sides may be inscribed.*

Ex. 438. The area of a circumscribed square is equal to twice the area of the inscribed square.

Ex. 439. The area of a circular ring is equal to that of a circle whose diameter is a chord of the outer circle tangent to the inner circle.

PROPOSITION XIV. PROBLEM.

469. *To inscribe a regular hexagon in a given circle.*

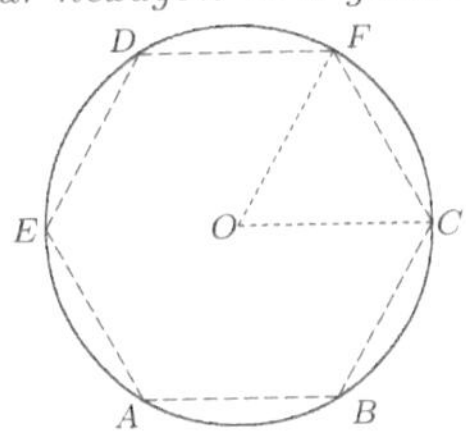

Let O be the centre of the given circle.

To inscribe a regular hexagon in the given circle.

From O draw any radius, as OC.

From C as a centre, with a radius equal to OC,

describe an arc intersecting the circumference at F.

Draw OF and CF.

Then CF is a side of the regular hexagon required.

Proof.

The OFC is equiangular, § 146

(*since it is equilateral by construction*).

Hence, the FOC is $\frac{1}{3}$ of 2 rt. s, or $\frac{1}{6}$ of 4 rt. s. § 136

the arc FC is $\frac{1}{6}$ of the circumference,

and the chord FC is a side of a regular inscribed hexagon.

Hence, to inscribe a regular hexagon apply the radius six times as a chord. Q.E.F.

470. COR. 1. *By joining the alternate vertices A, C, D, an equilateral triangle is inscribed in the circle.*

471. COR. 2. *By bisecting the arcs AB, BC, etc., a regular polygon of twelve sides may be inscribed in the circle; and, by continuing the process, regular polygons of twenty-four, forty-eight, etc., sides may be inscribed.*

Proposition XV. Problem.

472. *To inscribe a regular decagon in a given circle.*

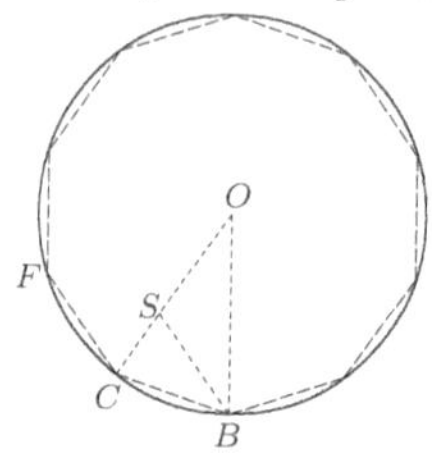

Let O be the centre of the given circle.

To inscribe a regular decagon in the given circle.

Draw any radius OC,

and divide it in extreme and mean ratio, so that OC shall

be to OS as OS is to SC. § 389

From C as a centre, with a radius equal to OS,

describe an arc intersecting the circumference at B.

Draw BC.

Then BC is a side of the regular decagon required.

Proof.

Draw BS and BO.

Now

$OC : OS = OS : SC,$ Const.

and

$BC = OS.$ Const.

$OC : BC = BC : SC.$

Moreover,

$OCB = SCB.$ Iden.

Hence, the $_{s}OCB$ and BCS are similar. § 357

But the OCB is isosceles. § 217

BCS, which is similar to the OCB, is isosceles,

and $CB = BS = SO.$ § 120

the SOB is isosceles, and the $O = SBO.$ § 145

But the ext. $CSB = \ O + \ SBO = 2 \ O$. § 137

Hence,

$$SCB = 2 \ O,$$

and

$$OBC = 2 \ O.$$

the sum of the $_s$ of the $OCB = 5 \ O = 2\text{rt.} \ _s$,

and

$$O = \tfrac{1}{5} \text{ of } 2 \text{ rt.} \ _s, \text{ or } \tfrac{1}{10} \text{ of } 4 \text{ rt.} \ _s.$$

Therefore, the arc BC is $\frac{1}{10}$ of the circumference,

and the chord BC is a side of a regular inscribed decagon.

Therefore, to inscribe a regular decagon, divide the radius internally in extreme and mean ratio, and apply the greater segment ten times as a chord. Q.E.F.

473. Cor. 1. *By joining the alternate vertices of a regular inscribed decagon, a regular pentagon is inscribed.*

474. Cor. 2. *By bisecting the arcs BC, CF, etc., a regular polygon of twenty sides may be inscribed in the circle; and, by continuing the process, regular polygons of forty, eighty, etc., sides may be inscribed.*

If R denotes the radius of a regular inscribed polygon, r the apothem, a one side, A an interior angle, and C the angle at the centre, show that

Ex. 440. In a regular inscribed triangle $a = R\sqrt{3}$, $r = \frac{1}{2}R$, $A = 60°$, $C = 120°$.

Ex. 441. In an inscribed square $a = R\sqrt{2}$, $r = \frac{1}{2}R\sqrt{2}$, $A = 90°$, $C = 90°$.

Ex. 442. In a regular inscribed hexagon $a = R$, $r = \frac{1}{2}R\sqrt{3}$, $A = 120°$, $C = 60°$.

Ex. 443. In a regular inscribed decagon

$$a = \frac{R(\sqrt{5}-1)}{2}, \ r = \frac{1}{4}R\sqrt{10+2\sqrt{5}}, \ A = 144°, \ C = 36°.$$

Proposition XVI. Problem.

475. *To inscribe in a given circle a regular pentedecagon, or polygon of fifteen sides.*

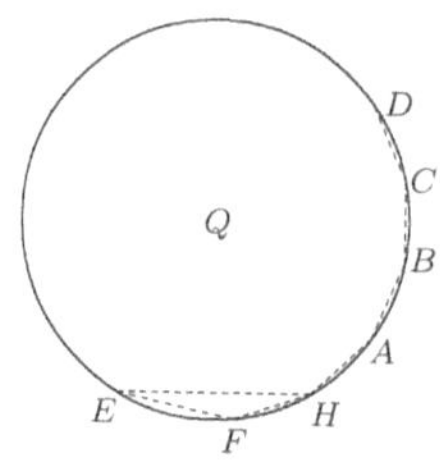

Let Q be the given circle.

To inscribe in Q a regular pentedecagon.

Draw EH equal to the radius of the circle,

and EF equal to a side of the regular inscribed decagon. § 472

Draw FH.

Then FH is a side of the regular pentedecagon required.

Proof.

The arc EH is $\frac{1}{6}$ of the circumference, § 469

and the arc EF is $\frac{1}{10}$ of the circumference. Const.

Hence, the arc FH is $\frac{1}{6} - \frac{1}{10}$, or $\frac{1}{15}$, of the circumference.

And the chord FH is a side of a regular inscribed pentedecagon.

By applying FH fifteen times as a chord, we have the polygon required. Q.E.F.

476. Cor. *By bisecting the arcs FH, HA, etc., a regular polygon of thirty sides may be inscribed; and, by continuing the process, regular polygons of sixty, one hundred twenty, etc., sides may be inscribed.*

PROPOSITION XVII. PROBLEM.

477. *To inscribe in a given circle a regular polygon similar to a given regular polygon.*

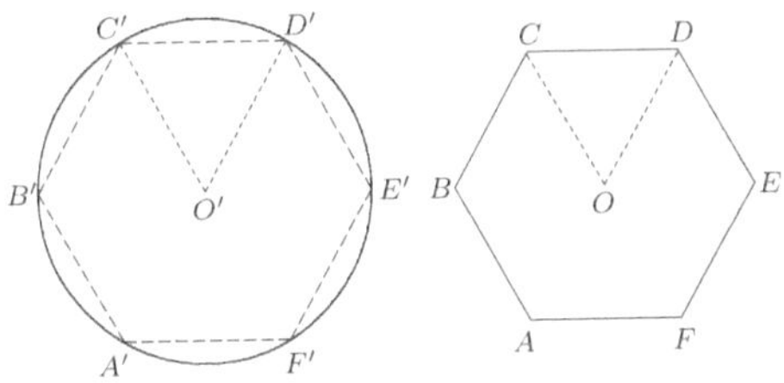

Let ABC etc. be the given regular polygon, and O' the centre of the given circle.

To inscribe in the circle a regular polygon similar to ABC etc.

From O, the centre of the given polygon,

draw OD and OC.

From O', the centre of the given circle,

draw $O'C'$ and $O'D'$,

making the O' equal to the O.

Draw $C'D'$.

Then $C'D'$ is a side of the regular polygon required.

Proof. Each polygon has as many sides as the O, or O', is contained times in 4 rt. s.

Therefore, the polygon $C'D'E'$ etc. is similar to the polygon CDE etc., § 445

(*two regular polygons of the same number of sides are similar*).

Q.E.F.

Ex. 444. The area of an inscribed regular octagon is equal to that of the rectangle whose sides are equal to the sides of the inscribed and the circumscribed squares.

PROPOSITION XVIII. PROBLEM.

478. *Given the side and the radius of a regular inscribed polygon, to find the side of the regular inscribed polygon of double the number of sides.*

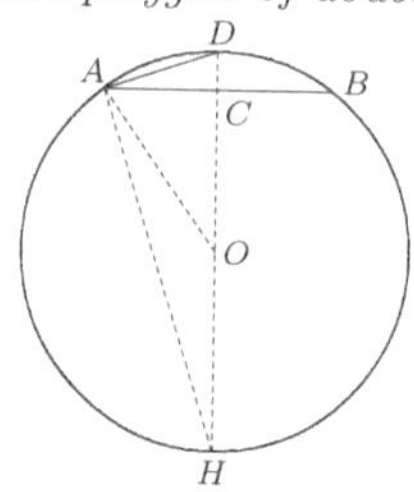

Let AB be a side of the regular inscribed polygon.

To find AD, a side of the regular inscribed polygon of double the number of sides.

Denote the radius by R, and AB by a.

From D draw DH through the centre O, and draw OA, AH.

DH is to AB at its middle point C. § 161

In the rt. OCA,

$$\overline{OC}^2 = R^2 - \tfrac{1}{4}a^2. \quad \text{§ 372}$$

Therefore,

$$OC = \sqrt{R^2 - \tfrac{1}{4}a^2},$$

and

$$DC = R - \sqrt{R^2 - \tfrac{1}{4}a^2}.$$

The DAH is a rt. . § 290

In the rt. DAH, $\overline{AD}^2 = DH \times DC$. § 367

But $DH = 2R$, and $DC = R - \sqrt{R^2 - \tfrac{1}{4}a^2}$.

$$AD = \sqrt{2R(R - \sqrt{R^2 - \tfrac{1}{4}a^2})}$$

$$= \sqrt{R(2R - \sqrt{4R^2 - a^2})}. \quad \text{Q.E.F.}$$

479. COR. *If $R = 1$, $AD = \sqrt{2 - \sqrt{4 - a^2}}$.*

PROPOSITION XIX. PROBLEM.

480. *To find the numerical value of the ratio of the circumference of a circle to its diameter.*

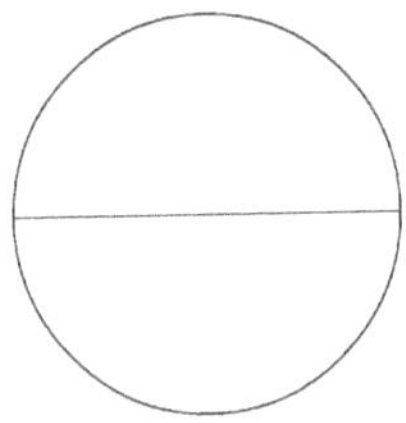

Let C be the circumference, when the radius is unity.

To find the numerical value of π.

By § 458, $2\pi R = C$. $\quad\quad \pi = \frac{1}{2}C$ when $R = 1$.

Let S_6 be the length of a side of a regular polygon of 6 sides, S_{12} of 12 sides, and so on.

If $R = 1$, by § 469, $S_6 = 1$ and by § 479 we have

Form of Computation.	Length of Side.	Length of Perimeter.
$S_{12} = \sqrt{2 - \sqrt{4 - 1^2}}$	0.51763809	6.21165708
$S_{24} = \sqrt{2 - \sqrt{4 - (0.51763809)^2}}$	0.26105238	6.26525722
$S_{48} = \sqrt{2 - \sqrt{4 - (0.26105238)^2}}$	0.13080626	6.27870041
$S_{96} = \sqrt{2 - \sqrt{4 - (0.13080626)^2}}$	0.06543817	6.28206396
$S_{192} = \sqrt{2 - \sqrt{4 - (0.06543817)^2}}$	0.03272346	6.28290510
$S_{384} = \sqrt{2 - \sqrt{4 - (0.03272346)^2}}$	0.01636228	6.28311544
$S_{768} = \sqrt{2 - \sqrt{4 - (0.01636228)^2}}$	0.00818121	6.28316941

$C = 6.28317$ approximately; that is, $\pi = 3.14159$ nearly. Q.E.F.

481. SCHOLIUM. π is incommensurable. We generally take

$$\pi = 3.1416, \text{ and } \frac{1}{\pi} = 0.31831.$$

MAXIMA AND MINIMA.

482. Def. Among geometrical magnitudes which satisfy given conditions, the *greatest* is called the **maximum**; and the *smallest* is called the **minimum**.

Thus, the diameter of a circle is the maximum among all chords; and the perpendicular is the minimum among all lines drawn to a given line from a given external point.

483. Def. **Isoperimetric** polygons are polygons which have equal perimeters.

Proposition XX. Theorem.

484. *Of all triangles having two given sides, that in which these sides include a right angle is the maximum.*

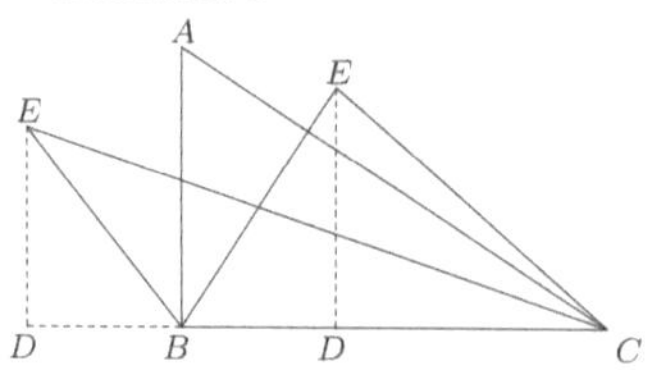

Let the triangles ABC and EBC have the sides AB and BC equal to EB and BC, respectively; and let the angle ABC be a right angle.

To prove that

$$ABC > \quad EBC.$$

Proof.

From E draw the altitude ED.

The ${}_sABC$ and EBC, having the same base, BC, are to each other as their altitudes AB and ED. § 405

Now

$$EB > ED. \qquad \text{§ 97}$$

But

$$EB = AB. \qquad \text{Hyp.}$$

$$AB > ED.$$

$$ABC > \quad EBC. \qquad \text{§ 405}$$

Q.E.D.

PROPOSITION XXI. THEOREM.

485. *Of all isoperimetric triangles having the same base the isosceles triangle is the maximum.*

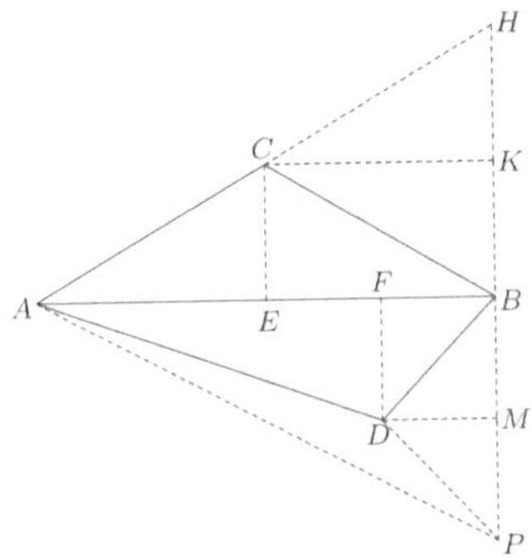

Let the $\triangle_s ACB$ and ADB have equal perimeters, and let AC and CB be equal, and AD and DB be unequal.

To prove that $ACB >$ ADB.

Proof.

Produce AC to H, making $CH = AC$; and draw HB.

Produce HB, take DP equal to DB, and draw AP.

Draw CE and DF to AB, and CK and $DM \parallel$ to AB.

The ABH is a right , for it may be inscribed in the semicircle whose centre is C and radius CA. § 290

ADP is not a straight line, for then the $_s DBA$ and DAB would be equal, being complements of the equal $_s DBM$ and DPM, respectively; and DA and DB would be equal (§ 147), which is contrary to the hypothesis. Hence,

$$AP < AD + DP, \quad < AD + DB, \quad < AC + CB, \quad < AH.$$

$$BH > BP. \qquad \text{§ 102}$$

$$CE (= \tfrac{1}{2} BH) > DF (= \tfrac{1}{2} BP). \qquad \text{Ax. 7}$$

Therefore,

$$ACB > \quad ADB. \qquad \text{§ 405}$$

Q.E.D.

PROPOSITION XXII. THEOREM.

486. *Of all polygons with sides all given but one, the maximum can be inscribed in a semicircle which has the undetermined side for its diameter.*

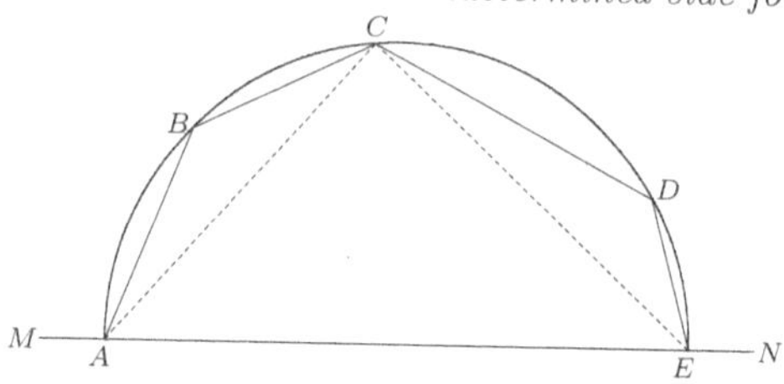

Let $ABCDE$ be the maximum of polygons with sides AB, BC, CD, DE, and the extremities A and E on the straight line MN.

To prove that $ABCDE$ can be inscribed in a semicircle.

Proof. From *any* vertex, as C, draw CA and CE.

The ACE must be the maximum of all ${}_s$ having the sides CA and CE, and the third side on MN; otherwise by increasing or diminishing the ACE, keeping the lengths of the sides CA and CE unchanged, but sliding the extremities A and E along the line MN, we could increase the ACE, while the rest of the polygon would remain unchanged; and therefore increase the polygon. But this is contrary to the hypothesis that the polygon is the maximum polygon.

Hence, the ACE is the maximum of ${}_s$ that have the sides CA and CE.

Therefore, the ACE is a right angle. § 484

Therefore, C lies on the semicircumference. § 290

Hence, *every* vertex lies on the circumference; that is, the maximum polygon can be inscribed in a semicircle having the undetermined side for a diameter. Q.E.D.

PROPOSITION XXIII. THEOREM.

487. *Of all polygons with given sides, that which can be inscribed in a circle is the maximum.*

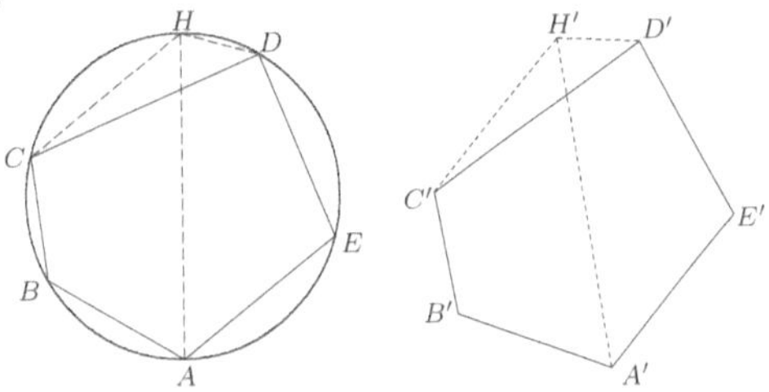

Let $ABCDE$ be a polygon inscribed in a circle, and $A'B'C'D'E'$ be a polygon, equilateral with respect to $ABCDE$, which cannot be inscribed in a circle.

To prove that that $ABCDE > A'B'C'D'E'$.

Proof.

Draw the diameter AH, and draw CH and DH.

Upon $C'D'$ construct the $C'H'D' =$ CHD, and draw $A'H'$.

Since, by hypothesis, a ⊙ cannot pass through *all* the vertices of $A'B'C'D'E'$, *one* or *both* of the parts $ABCH$, $AEDH$ must be greater than the corresponding part of $A'B'C'H'D'E'$. § 486

If either of these parts is *not greater than* its corresponding part, it is equal to it, § 486

(*for $ABCH$ and $AEDH$ are the maxima of polygons that have sides equal to AB, BC, CH, and AE, ED, DH, respectively, and the remaining side undetermined*).

$$ABCHDE > A'B'C'H'D'E'. \qquad \text{Ax. 4}$$

Take away from the two figures the equal ${}_sCHD$ and $C'H'D'$.

Then

$$ABCDE > A'B'C'D'E'. \qquad \text{Ax. 5}$$

Q.E.D.

PROPOSITION XXIV. THEOREM.

488. *Of isoperimetric polygons of the same number of sides, the maximum is equilateral.*

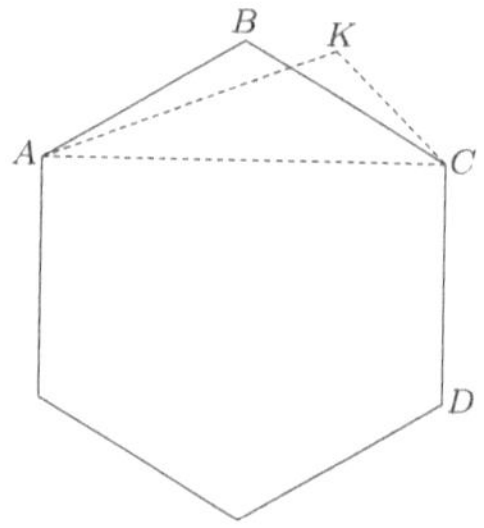

Let $ABCD$ etc. be the maximum of isoperimetric polygons of any given number of sides.

To prove that AB, BC, CD, etc., are equal.

Proof.

Draw AC.

The ABC must be the maximum of all the s which are formed upon AC with a perimeter equal to that of ABC.

Otherwise a greater AKC could be substituted for ABC, without changing the perimeter of the polygon.

But this is inconsistent with the hypothesis that the polygon $ABCD$ etc. is the maximum polygon.

the ABC is isosceles. § 485

$AB = BC$.

In like manner it may be proved that $BC = CD$, etc. Q.E.D.

489. COR. *The maximum of isoperimetric polygons of the same number of sides is a regular polygon.*

For the maximum polygon is equilateral (§ 488), and can be inscribed in a circle (§ 487), and is, therefore, regular. § 430

Proposition XXV. Theorem.

490. *Of isoperimetric regular polygons, that which has the greatest number of sides is the maximum.*

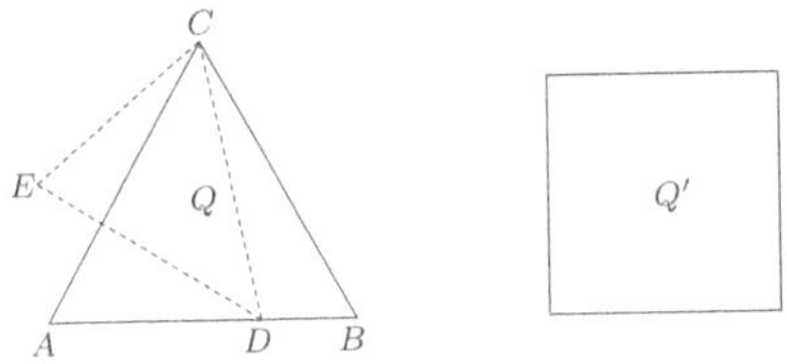

Let Q be a regular polygon of three sides, and Q' a regular polygon of four sides, and let the two polygons have equal perimeters.

To prove that Q' is greater than Q.

Proof. Draw CD from C to any point in AB.

Invert the CDA and place it in the position DCE, letting D fall at C, C at D, and A at E.

The polygon $DBCE$ is an irregular polygon of four sides, which by construction has the same perimeter as Q', and the same area as Q.

Then the irregular polygon $DBCE$ of four sides is less than the isoperimetric regular polygon Q' of four sides. § 489

In like manner it may be shown that Q' is less than an isoperimetric regular polygon of five sides, and so on. Q.E.D.

Ex. 445. Of all equivalent parallelograms that have equal bases, the rectangle has the minimum perimeter.

Ex. 446. Of all equivalent rectangles, the square has the minimum perimeter.

Ex. 447. Of all triangles that have the same base and the same altitude, the isosceles has the minimum perimeter.

Ex. 448. Of all triangles that can be inscribed in a given circle, the equilateral is the maximum and has the maximum perimeter.

PROPOSITION XXVI. THEOREM.

491. *Of regular polygons having a given area, that which has the greatest number of sides has the least perimeter.*

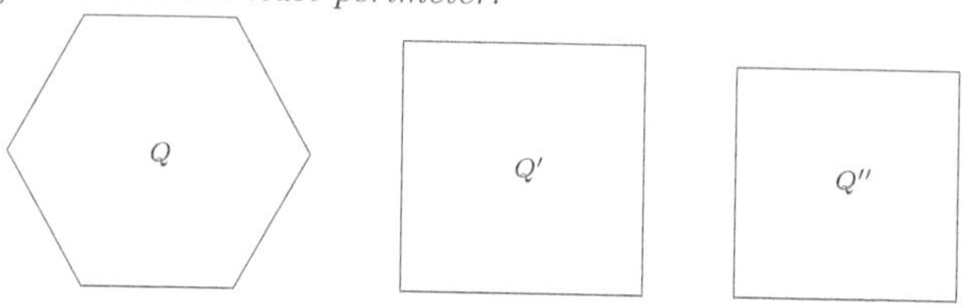

Let Q and Q' be regular polygons having the same area, and let Q' have the greater number of sides.

To prove

$$\text{the perimeter of } Q > \text{the perimeter of } Q'.$$

Proof. Let Q'' be a regular polygon having the same perimeter as Q', and the same number of sides as Q.

Then

$$Q' > Q'' \qquad \S\ 490$$

(*of isoperimetric regular polygons, that which has the greatest number of sides is the maximum*).

But

$$Q \quad Q'. \qquad \text{Hyp.}$$

$$Q > Q''.$$

$$\text{the perimeter of } Q > \text{the perimeter of } Q''.$$

$$\text{But the perimeter of } Q' = \text{the perimeter of } Q''. \qquad \text{Hyp.}$$

$$\text{the perimeter of } Q > \text{the perimeter of } Q'. \qquad \text{Q.E.D.}$$

Ex. 449. To inscribe in a semicircle the maximum rectangle.

Ex. 450. Of all polygons of a given number of sides which may be inscribed in a given circle, that which is regular has the maximum area and the maximum perimeter.

Ex. 451. Of all polygons of a given number of sides which may be circumscribed about a given circle, that which is regular has the minimum area and the minimum perimeter.

THEOREMS.

Ex. 452. Every equilateral polygon circumscribed about a circle is regular if it has an *odd* number of sides.

Ex. 453. Every equiangular polygon inscribed in a circle is regular if it has an *odd* number of sides.

Ex. 454. Every equiangular polygon circumscribed about a circle is regular.

Ex. 455. The side of a circumscribed equilateral triangle is equal to twice the side of the similar inscribed triangle.

Ex. 456. The apothem of an inscribed regular hexagon is equal to half the side of the inscribed equilateral triangle.

Ex. 457. The area of an inscribed regular hexagon is three fourths of the area of the circumscribed regular hexagon.

Ex. 458. The area of an inscribed regular hexagon is the mean proportional between the areas of the inscribed and the circumscribed equilateral triangles.

Ex. 459. The square of the side of an inscribed equilateral triangle is equal to three times the square of a side of the inscribed regular hexagon.

Ex. 460. The area of an inscribed equilateral triangle is equal to half the area of the inscribed regular hexagon.

Ex. 461. The square of the side of an inscribed equilateral triangle is equal to the sum of the squares of the sides of the inscribed square and of the inscribed regular hexagon.

Ex. 462. The square of the side of an inscribed regular pentagon is equal to the sum of the squares of the radius of the circle and the side of the inscribed regular decagon.

If R denotes the radius of a circle, and a one side of an inscribed regular polygon, show that:

Ex. 463. In a regular pentagon, $a = \frac{1}{2}R\sqrt{10 - 2\sqrt{5}}$.

Ex. 464. In a regular octagon, $a = R\sqrt{2 - \sqrt{2}}$.

Ex. 465. In a regular dodecagon, $a = R\sqrt{2 - \sqrt{3}}$.

Ex. 466. If two diagonals of a regular pentagon intersect, the longer segment of each is equal to a side of the pentagon.

Ex. 467. The apothem of an inscribed regular pentagon is equal to half the sum of the radius of the circle and the side of the inscribed regular decagon.

Ex. 468. The side of an inscribed regular pentagon is equal to the hypotenuse of the right triangle which has for legs the radius of the circle and the side of the inscribed regular decagon.

Ex. 469. The radius of an inscribed regular polygon is the mean proportional between its apothem and the radius of the similar circumscribed regular polygon.

Ex. 470. If squares are constructed outwardly upon the six sides of a regular hexagon, the exterior vertices of these squares are the vertices of a regular dodecagon.

Ex. 471. If the alternate vertices of a regular hexagon are joined by straight lines, show that another regular hexagon is thereby formed. Find the ratio of the areas of these two hexagons.

Ex. 472. If on the legs of a right triangle as diameters semicircles are described external to the triangle, and from the whole figure a semicircle on the hypotenuse is subtracted, the remaining figure is equivalent to the given right triangle.

Ex. 473. The star-shaped polygon, formed by producing the sides of a regular hexagon, is equivalent to twice the given hexagon.

Ex. 474. The sum of the perpendiculars drawn to the sides of a regular polygon from any point within the polygon is equal to the apothem multiplied by the number of sides.

Ex. 475. If two chords of a circle are perpendicular to each other, the sum of the four circles described on the four segments as diameters is equivalent to the given circle.

Ex. 476. If the diameter of a circle is divided into any two segments, and upon these segments as diameters semicircumferences are described upon opposite sides of the diameter, these semicircumferences divide the circle into two parts which have the same ratio as the two segments of the diameter.

Ex. 477. The diagonals that join any vertex of a regular polygon to all the vertices not adjacent divide the angle at that vertex into as many equal parts less two as the polygon has sides.

PROBLEMS OF CONSTRUCTION.

Ex. 478. To circumscribe an equilateral triangle about a given circle.

Ex. 479. To circumscribe a square about a given circle.

Ex. 480. To circumscribe a regular hexagon about a given circle.

Ex. 481. To circumscribe a regular octagon about a given circle.

Ex. 482. To circumscribe a regular pentagon about a given circle.

Ex. 483. To draw through a given point a line so as to divide a given circumference into two parts having the ratio 3 : 7.

Ex. 484. To construct a circumference equal to the sum of two given circumferences.

Ex. 485. To construct a circumference equal to the difference of two given circumferences.

Ex. 486. To construct a circle equivalent to the sum of two given circles.

Ex. 487. To construct a circle equivalent to the difference of two given circles.

Ex. 488. To construct a circle equivalent to three times a given circle.

Ex. 489. To construct a circle equivalent to three fourths of a given circle.

Ex. 490. To construct a circle whose ratio to a given circle shall be equal to the given ratio $m : n$.

Ex. 491. To divide a given circle by a concentric circumference into two equivalent parts.

Ex. 492. To divide a given circle by concentric circumferences into five equivalent parts.

Ex. 493. To construct an angle of 18°; of 36°; of 9°.

Ex. 494. To construct an angle of12°; of 24°; of 6°.

To construct with a side of a given length:

Ex. 495. An equilateral triangle.

Ex. 496. A square.

Ex. 497. A regular hexagon.

Ex. 498. A regular octagon.

Ex. 499. A regular pentagon.

Ex. 500. A regular decagon.

Ex. 501. A regular dodecagon.

Ex. 502. A regular pentedecagon.

PROBLEMS OF COMPUTATION.

Ex. 503. Find the area of a circle whose radius is 12 inches.

Ex. 504. Find the circumference and the area of a circle whose diameter is 8 feet.

Ex. 505. A regular pentagon is inscribed in a circle whose radius is R. If the length of a side is a, find the apothem.

Ex. 506. A regular polygon is inscribed in a circle whose radius is R. If the length of a side is a, show that the apothem is $\frac{1}{2}\sqrt{R^2 - a^2}$.

Ex. 507. Find the area of a regular decagon inscribed in a circle whose radius is 16 inches.

Ex. 508. Find the side of a regular dodecagon inscribed in a circle whose radius is 20 inches.

Ex. 509. Find the perimeter of a regular pentagon inscribed in a circle whose radius is 25 feet.

Ex. 510. The length of each side of a park in the shape of a regular decagon is 100 yards. Find the area of the park.

Ex. 511. Find the cost, at \$2 per yard, of building a wall around a cemetery in the shape of a regular hexagon, that contains 16,627.84 square yards.

Ex. 512. The side of an inscribed regular polygon of n sides is 16 feet. Find the side of an inscribed regular polygon of $2n$ sides.

Ex. 513. If the radius of a circle is R, and the side of an inscribed regular polygon is a, show that the side of the similar circumscribed regular polygon is $\frac{2aR}{\sqrt{4R^2 - a^2}}$.

Ex. 514. What is the width of the circular ring between two concentric circumferences whose lengths are 650 feet and 425 feet?

Ex. 515. Find the angle subtended at the centre by an arc 5 feet 10 inches long, if the radius of the circle is 9 feet 4 inches.

Ex. 516. The chord of a segment is 10 feet, and the radius of the circle is 16 feet. Find the area of the segment.

Ex. 517. Find the area of a sector, if the angle at the centre is 20°, and the radius of the circle is 20 inches.

Ex. 518. The chord of half an arc is 12 feet, and the radius of the circle is 18 feet. Find the height of the segment subtended by the whole arc.

Ex. 519. Find the side of a square which is equivalent to a circle whose diameter is 35 feet.

Ex. 520. The diameter of a circle is 15 feet. Find the diameter of a circle twice as large. Three times as large.

Ex. 521. Find the radii of the concentric circumferences that divide a circle 11 inches in diameter into five equivalent parts.

Ex. 522. The perimeter of a regular hexagon is 840 feet, and that of a regular octagon is the same. By how many square feet is the octagon larger than the hexagon?

Ex. 523. The diameter of a bicycle wheel is 28 inches. How many revolutions does the wheel make in going 10 miles?

Ex. 524. Find the diameter of a carriage wheel that makes 264 revolutions in going half a mile.

Ex. 525. The sides of three regular octagons are 6 feet, 7 feet, 8 feet, respectively. Find the side of a regular octagon equivalent to the sum of the three given octagons.

Ex. 526. A circular pond 100 yards in diameter is surrounded by a walk 10 feet wide. Find the area of the walk.

Ex. 527. The span (chord) of a bridge in the form of a circular arc is 120 feet, and the highest point of the arch is 15 feet above the piers. Find the radius of the arc.

Ex. 528. Three equal circles are described each tangent to the other two. If the common radius is R, find the area contained between the circles.

Ex. 529. Given p, P, the perimeters of regular polygons of n sides inscribed in and circumscribed about a given circle. Find p', P', the perimeters of regular polygons of $2n$ sides inscribed in and circumscribed about the given circle.

Ex. 530. Given the radius R, and the apothem r of an inscribed regular polygon of n sides. Find the radius R' and the apothem r' of an isoperimetrical regular polygon of $2n$ sides.

MISCELLANEOUS EXERCISES.

THEOREMS.

Ex. 531. If two adjacent angles of a quadrilateral are right angles, the bisectors of the other two angles are perpendicular.

Ex. 532. If two opposite angles of a quadrilateral are right angles, the bisectors of the other two angles are parallel.

Ex. 533. The two lines that join the middle points of the opposite sides of a quadrilateral bisect each other.

Ex. 534. The line that joins the feet of the perpendiculars dropped from the extremities of the base of an isosceles triangle to the opposite sides is parallel to the base.

Ex. 535. If AD bisects the angle A of a triangle ABC, and BD bisects the exterior angle CBF, then angle ADB equals one half angle ACB.

Ex. 536. The sum of the acute angles at the vertices of a pentagram (five-pointed star) is equal to two right angles.

Ex. 537. The altitudes AD, BE, CF of the triangle ABC bisect the angles of the triangle DEF.

Circles with AB, BC, AC as diameters will pass through E and D, E and F, D and F, respectively.

Ex. 538. The segments of any straight line intercepted between the circumferences of two concentric circles are equal.

Ex. 539. If a circle is circumscribed about any triangle, the feet of the perpendiculars dropped from any point in the circumference to the sides of the triangle lie in one straight line.

Ex. 540. Two circles are tangent internally at P, and a chord AB of the larger circle touches the smaller circle at C. Prove that PC bisects the angle APB.

Ex. 541. The diagonals of a trapezoid divide each other into segments which are proportional.

Ex. 542. If through a point P in the circumference of a circle two chords are drawn, the chords and the segments between P and a chord parallel to the tangent at P are reciprocally proportional.

Ex. 543. The perpendiculars from two vertices of a triangle upon the opposite sides divide each other into segments reciprocally proportional.

Ex. 544. The perpendicular from any point of a circumference upon a chord is the mean proportional between the perpendiculars from the same point upon the tangents drawn at the extremities of the chord.

Ex. 545. In an isosceles right triangle either leg is the mean proportional between the hypotenuse and the perpendicular upon it from the vertex of the right angle.

Ex. 546. If two circles intersect in the points A and B, and through A any secant CAD is drawn limited by the circumferences at C and D, the straight lines BC, BD are to each other as the diameters of the circles.

Ex. 547. The area of a triangle is equal to half the product of its perimeter by the radius of the inscribed circle.

Ex. 548. The perimeter of a triangle is to one side as the perpendicular from the opposite vertex is to the radius of the inscribed circle.

Ex. 549. If three straight lines AA', BB', CC', drawn from the vertices of a triangle ABC to the opposite sides, pass through a common point O within the triangle, then

$$\frac{OA'}{AA'} + \frac{OB'}{BB'} + \frac{OC'}{CC'} = 1.$$

Ex. 550. ABC is a triangle, M the middle point of AB, P any point in AB between A and M. If MD is drawn parallel to PC, meeting BC at D, the triangle BPD is equivalent to half the triangle ABC.

Ex. 551. Two diagonals of a regular pentagon, not drawn from a common vertex, divide each other in extreme and mean ratio.

Ex. 552. If all the diagonals of a regular pentagon are drawn, another regular pentagon is thereby formed.

Ex. 553. The area of an inscribed regular dodecagon is equal to three times the square of the radius.

Ex. 554. The area of a square inscribed in a semicircle is equal to two fifths the area of the square inscribed in the circle.

Ex. 555. The area of a circle is greater than the area of any polygon of equal perimeter.

Ex. 556. The circumference of a circle is less than the perimeter of any polygon of equal area.

PROBLEMS OF LOCI.

Ex. 557. Find the locus of the centre of the circle inscribed in a triangle that has a given base and a given angle at the vertex.

Ex. 558. Find the locus of the intersection of the altitudes of a triangle that has a given base and a given angle at the vertex.

Ex. 559. Find the locus of the extremity of a tangent to a given circle, if the length of the tangent is equal to a given line.

Ex. 560. Find the locus of a point, tangents drawn from which to a given circle form a given angle.

Ex. 561. Find the locus of the middle point of a line drawn from a given point to a given straight line.

Ex. 562. Find the locus of the vertex of a triangle that has a given base and a given altitude.

Ex. 563. Find the locus of a point the sum of whose distances from two given parallel lines is equal to a given length.

Ex. 564. Find the locus of a point the difference of whose distances from two given parallel lines is equal to a given length.

Ex. 565. Find the locus of a point the sum of whose distances from two given intersecting lines is equal to a given length.

Ex. 566. Find the locus of a point the difference of whose distances from two given intersecting lines is equal to a given length.

Ex. 567. Find the locus of a point whose distances from two given points are in the given ratio $m : n$.

Ex. 568. Find the locus of a point whose distances from two given parallel lines are in the given ratio $m : n$.

Ex. 569. Find the locus of a point whose distances from two given intersecting lines are in the given ratio $m : n$.

Ex. 570. Find the locus of a point the sum of the squares of whose distances from two given points is constant.

Ex. 571. Find the locus of a point the difference of the squares of whose distances from two given points is constant.

Ex. 572. Find the locus of the vertex of a triangle that has a given base and the other two sides in the given ratio $m : n$.

PROBLEMS OF CONSTRUCTION.

Ex. 573. To divide a given trapezoid into two equivalent parts by a line parallel to the bases.

Ex. 574. To divide a given trapezoid into two equivalent parts by a line through a given point in one of the bases.

Ex. 575. To construct a regular pentagon, given one of the diagonals.

Ex. 576. To divide a given straight line into two segments such that their product shall be the maximum.

Ex. 577. To find a point in a semicircumference such that the sum of its distances from the extremities of the diameter shall be the maximum.

Ex. 578. To draw a common secant to two given circles exterior to each other such that the intercepted chords shall have the given lengths a, b.

Ex. 579. To draw through one of the points of intersection of two intersecting circles a common secant which shall have a given length.

Ex. 580. To construct an isosceles triangle, given the altitude and one of the equal base angles.

Ex. 581. To construct an equilateral triangle, given the altitude.

Ex. 582. To construct a right triangle, given the radius of the inscribed circle and the difference of the acute angles.

Ex. 583. To construct an equilateral triangle so that its vertices shall lie in three given parallel lines.

Ex. 584. To draw a line from a given point to a given straight line which shall be to the perpendicular from the given point as $m : n$.

Ex. 585. To find a point within a given triangle such that the perpendiculars from the point to the three sides shall be as the numbers m, n, p.

Ex. 586. To draw a straight line equidistant from three given points.

Ex. 587. To draw a tangent to a given circle such that the segment intercepted between the point of contact and a given straight line shall have a given length.

Ex. 588. To inscribe a straight line of a given length between two given circumferences and parallel to a given straight line.

Ex. 589. To draw through a given point a straight line so that its distances from two other given points shall be in a given ratio.

Ex. 590. To construct a square equivalent to the sum of a given triangle and a given parallelogram.

Ex. 591. To construct a rectangle having the difference of its base and altitude equal to a given line, and its area equivalent to the sum of a given triangle and a given pentagon.

Ex. 592. To construct a pentagon similar to a given pentagon and equivalent to a given trapezoid.

Ex. 593. To find a point whose distances from three given straight lines shall be as the numbers m, n, p.

Ex. 594. Given an angle and two points P and P' between the sides of the angle. To find the shortest path from P to P' that shall touch both sides of the angle.

Ex. 595. To construct a triangle, given its angles and its area.

Ex. 596. To transform a given triangle into a triangle similar to another given triangle.

Ex. 597. Given three points A, B, C. To find a fourth point P such that the areas of the triangles APB, APC, BPC shall be equal.

Ex. 598. To construct a triangle, given its base, the ratio of the other sides, and the angle included by them.

Ex. 599. To divide a given circle into n equivalent parts by concentric circumferences.

Ex. 600. In a given equilateral triangle to inscribe three equal circles tangent to each other, each circle tangent to two sides of the triangle.

Ex. 601. Given an angle and a point P between the sides of the angle. To draw through P a straight line that shall form with the sides of the angle a triangle with the perimeter equal to a given length a.

Ex. 602. In a given square to inscribe four equal circles, so that each circle shall be tangent to two of the others and also tangent to two sides of the square.

Ex. 603. In a given square to inscribe four equal circles, so that each circle shall be tangent to two of the others and also tangent to one side of the square.

TABLE OF FORMULAS.

PLANE FIGURES.

NOTATION.

P = perimeter.
h = altitude.
b = lower base.
b' = upper base.
R = radius of circle.
D = diameter of circle.
C = circumference of circle.
r = apothem of regular polygon.
a, b, c = sides of triangle.
$s = \frac{1}{2}(a + b + c)$.
p = perpendicular of triangle.
m, n = segments of third side of triangle adjacent to sides b and a, respectively.
S = area.
$\pi = 3.1416$.

FORMULAS.

Line Values.

		PAGE
Right triangle,	$b^2 = c \times m;\ a^2 = c \times n$	197
	$p^2 = m \times n$	197
	$b^2 : a^2 = m : n$	198
	$b^2 : c^2 :: m : c$	198
	$a^2 + b^2 = c^2$	199
Any triangle,	$a^2 = b^2 + c^2 \pm 2c \times m$	200,201
Altitude of triangle on side a,	$h = \frac{2}{a}\sqrt{s(s-a)(s-b)(s-c)}$	219
Median of triangle on side a,	$m = \frac{1}{2}\sqrt{2(b^2+c^2)-a^2}$	220
Bisector of triangle on side a,	$t = \frac{2}{b+c}\sqrt{bcs(s-a)}$	221
Radius of circumscribed circle,	$R = \frac{abc}{4\sqrt{s(s-a)(s-b)(s-c)}}$	222
Circumference of circle,	$C = 2\pi R$	270
" "	$C = \pi D$	270

Areas.

Rectangle,	$S = b \times h$	229
Square,	$S = b^2$	229
Parallelogram,	$S = b \times h$	230
Triangle,	$S = \frac{1}{2}b \times h$	231
"	$S = \sqrt{s(s-a)(s-b)(s-c)}$	254
"	$S = \frac{abc}{4R}$	254
Equilateral triangle,	$S = \frac{a^2}{4}\sqrt{3}$	253
Trapezoid,	$S = \frac{1}{2}h(b+b')$	232
Regular polygon,	$S = \frac{1}{2}r \times P$	270
Circle,	$S = \frac{1}{2}R \times C$	272
"	$S = \pi R^2$	272
Sector,	$S = \frac{1}{2}R \times \text{arc}$	271

INDEX.

9 789393 971449

Printed at Repro India Ltd.